THERMAL PHYSICS

DATE DUE

THERMAL PHYSICS

An introduction to thermodynamics, statistical mechanics, and kinetic theory

Second Edition

P. C. Riedi

Department of Physics
University of St. Andrews

Oxford   New York   Tokyo
OXFORD UNIVERSITY PRESS
1988

Oxford University Press, Walton Street, Oxford OX2 6DP
Oxford New York Toronto
Delhi Bombay Calcutta Madras Karachi
Petaling Jaya Singapore Hong Kong Tokyo
Nairobi Dar es Salaam Cape Town
Melbourne Auckland
and associated companies in
Berlin Ibadan

Oxford is a trade mark of Oxford University Press

Published in the United States
by Oxford University Press, New York

The first edition of this book was published by the Macmillan Press Ltd in 1976
Second edition 1988

British Library Cataloguing in Publication Data
Riedi, P. C.
Thermal physics : an introduction to
thermodynamics, statistical mechanics, and
kinetic theory.—2nd ed.
1. Thermodynamics
I. Title
536'.7
ISBN   0–19–853922–3
ISBN   0–19–851992–3 Pbk

Library of Congress Cataloging in Publication Data

Riedi, P. C.
Thermal physics.
Bibliography: p. 335 includes index.
1. Thermodynamics.   2. Statistical mechanics.
3. Gases, Kinetic theory of.   I. Title.
QC311.R52   1988   536'.7   88–5362
ISBN   0–19–853922–3
ISBN   0–19–851992–3 (pbk.)

Typeset by Macmillan India Limited, Bangalore 560 025
Printed in Great Britain
at the University Printing House, Oxford
by David Stanford
Printer to the University

Preface to the Second Edition

I have taken the opportunity provided by a new edition to reorganize some parts of the book and to bring the more advanced material up to date. The number of exercises has been greatly increased to provide both straight-forward practice in the use of basic concepts and examples of the application of these ideas in modern science and technology. Some of the new exercises are designed to form the basis for an essay topic or term paper and have references provided. Three appendices have been added to the first edition. That to Chapter 3 provides the material required to introduce the second law of thermodynamics in the traditional manner if the approach given in the body of the chapter is thought too abstract for a first course in thermodyn-amics. The appendices to Chapters 4 and 7 introduce the thermodynamics and statistical mechanics of systems with a variable number of particles. This material may be omitted on a first reading.

I am indebted to Professor D. V. Osborne and Professor E. Braun for helpful comments on the first edition, to various colleagues for reading parts of the new material, and to Miss K. Lumsden and Miss L. MacPherson for typing the new sections of the manuscript.

St. Andrews
June 1987

P. C. R.

Preface to the First Edition

The number of lectures devoted to traditional subjects such as thermodynamics has decreased recently in many honours degree courses owing to the commendable desire to introduce current research topics to undergraduates. The leisurely discussion of thermodynamics given in the standard undergraduate texts is now rather out of proportion to the time that the student is prepared to devote to the subject. One method of saving time is to teach a course purely from the atomic view of matter and in some way to 'derive' the laws of thermodynamics from the results of statistical mechanics. This approach is also claimed to be more likely to arouse the interest of students already familiar with elementary atomic physics but has the great disadvantage of presenting thermodynamics as a trivial and dependent subject rather than as one of the greatest achievements of physics.

There is no doubt that much can be learnt by a judicious mixture of the macroscopic and microscopic approaches and so this book covers both thermodynamics and statistical mechanics. However, they are first introduced separately and then their strengths and weaknesses are further explored by examining a number of selected topics from both points of view. A chapter is also devoted to the kinetic theory of the transport properties of gases, partly because of the importance of the results in such fields as vacuum physics and partly to emphasize the great increase in difficulty associated with the study of systems away from thermal equilibrium.

The main objections to the teaching of thermodynamics without the introduction of statistical mechanics have always seemed to me to centre on the tortuous approach to the second law of thermodynamics given in most elementary textbooks. An engineer may find proofs based on hypothetical engines driving each other backwards and forwards fascinating, but to many physics students these seem rather special phenomena on which to base a general law of science. The physical *need* for a second law of thermodynamics is therefore discussed in some detail in Chapters 2 and 3 and then the mathematical theorem of Carathéodory stated. The distinction between this theorem and the statement of the second law of thermodynamics in the form given by Carathéodory is then carefully explained. Once the concept of an integrating factor for the first law of thermodynamics has been established the whole subject can be developed in a logical fashion and heat engines dealt with *en passant*.

The wide application of the methods of thermodynamics and statistical mechanics in modern research is demonstrated in Chapter 10 where such varied subjects as the 3 K stellar background radiation, Pomeranchuk cooling of ^{3}He to within a few thousandths of a degree of absolute zero, the thermodynamic inequalities at phase transitions, and negative temperature are treated, and references are given to fuller accounts in accessible journals.

It is hoped that these modern examples will both arouse the interest of the student and impress upon him the continuing importance of a subject whose origins lie in the nineteenth century but which will always occupy a central position in physics.

I am indebted to various colleagues for comments on parts of the manuscript to the copyright holders for permission to use certain figures and to Mrs M. Gray for typing the manuscript.

St. Andrews   P. C. R.
1976

Contents

PART IV APPLICATIONS OF THERMODYNAMICS AND
STATISTICAL MECHANICS

Symbols

a	given radius; van der Waals constant
b	van der Waals constant, constant
B_V, B_P	virial coefficients
B_T	isothermal bulk modulus
$C_V(C_P)$	heat capacity at constant volume (pressure); virial coefficients
$c_V(c_P)$	specific heat per molecule at constant volume (pressure)
c	speed of light
c_s	speed of sound
d	diameter
E	energy of whole system
$\mathscr{F}$	tension
$G(g)$	Gibbs free energy (per unit mass)
$H(h)$	Enthalpy (per unit mass)
I	nuclear spin
J	rotational quantum number
$\mathbf{k}(k)$	wave vector (magnitude)
k	Boltzmann constant
l	latent heat
m	mass of one molecule
m_0	mass of system
$\mathscr{M}$	total magnetic moment
M	molar mass, magnetization
N	number of molecules
N_A	Avogadro constant
n	number of molecules per unit volume
$\mathbf{p}(p)$	momentum (magnitude)
P	pressure
Q	quantity of heat
$R(R_m)$	gas constant per mole (per unit mass)
r	position, integer (subscript)
$S(s)$	entropy (per unit mass); spin of a particle
T	absolute temperature
$U(u)$	internal energy (per unit volume)

V	volume
v	speed
$\boldsymbol{v}$	velocity
W	work; number of microstates
$Z(z)$	partition function of system (of one particle)
$\mathscr{Z}$	grand partition function
α	coefficient of expansion, Lagrange undetermined multiplier; critical index
β	$1/kT$ (Lagrange undetermined multiplier); critical index
γ	C_P/C_V; Lagrange undetermined multiplier; coefficient of electronic specific heat, critical index
$\Gamma(n)$	gamma function (Appendix II)
ε	individual particle energy
$\zeta(z)$	Riemann zeta function (Appendix II)
η	coefficient of viscosity
θ	temperature; angle
κ	thermal conductivity
κ_S	adiabatic compressibility
κ_T	isothermal compressibility
λ	mean free path; wavelength
μ	magnetic moment; Lagrange undetermined multiplier, chemical potential
v	frequency
ρ	density
σ	scattering cross-section; Stefan–Boltzmann constant
τ	relaxation time
ϕ	angle; integrating factor
χ	susceptibility
Ψ, ψ	wavefunction

A note on units

SI units (kilogram-metre-second) are largely used in this book but other units are still so common in physics that it is essential for the student to know of their existence. Approximate values of the fundamental constants in SI units are given in Appendix V for use in calculations.

The SI unit of pressure is the Pascal (Pa) which is equivalent to $1 \, \text{N} \, \text{m}^{-2}$. Although this unit is becoming more popular, in research publications it is still common for high pressures to be measured in kbar or atmospheres (atm) and for the pressure in vacuum systems to be given in mm of mercury (torr). The relationship between these quantities is

760 mm of mercury $= 1$ atm $\doteq 1.01325 \times 10^5$ Pa
  1 mm of mercury $= 1$ torr $= 133.322$ Pa
                    1 bar $= 1 \times 10^5$ Pa (exact).

One other non-standard unit still in common use is the Angstrom (Å) as a unit of length in atomic physics:

$1 \, \text{Å} = 1 \times 10^{-10} \, \text{m} = 0.1 \, \text{nm}$.

Magnetic fields in free space will be written $\mu_0 H^*$ so that the basic unit is the tesla (T) which has a simple relation to the c.g.s. system

$1 \, \text{T} = 10^4$ Gauss (G).

To the student

Thermal physics is the study of the properties of systems containing a large number of atoms. In this sense it therefore covers nearly all physics although it will be seen that most of this book is concerned with the special case of systems in thermal equilibrium. The concepts of thermal physics, perhaps more than any other branch of physics, are most easily grasped in detail once a certain breadth of experience has been obtained. This short book—the essentials of the subject are developed in 167 pages—is therefore designed to provide a broad foundation for the study of the more difficult concepts and applications of thermal physics and does not attempt to provide completely rigorous proofs of every point which is discussed. Consult the reading list at the back of the book when you find a point which does not seem to be dealt with to your satisfaction or when you wish to increase your knowledge of some aspect of the subject. The mathematics required for the book is revised in Appendixes I–III.

At the end of each chapter will be found a set of exercises. It is essential that the reader attempt these exercises—some of the results of which are used in the text—and is able to understand the answers given at the end of the book, although a fuller understanding of, say, Chapter 3 may well come after reading Chapter 7.

The final chapter is designed to show the techniques of thermal physics at work. The selection of topics includes some of the most interesting recent developments in physics and astronomy as well as a number of more traditional topics which were too important to leave out.

To summarize: try to see Chapters 2–7 *as a whole*, work through the exercises and study the solutions, and use the book list to extend your knowledge.

1

Introduction

The purpose of this chapter is to revise certain concepts which the reader has probably already met and to define some technical terms which will be required in later chapters. The theme of the chapter is the essential difference between the methods which have been developed to handle small numbers of particles (mechanics) and those required for the very large number of particles (atoms or molecules) contained in a typical laboratory sample of a substance. All the material in this chapter is covered in more detail later in the book but an overview may be helpful before beginning a systematic study of thermal physics.

1.1 Preliminary survey

Classical mechanics is concerned with the motion of a *small* number of bodies, such as the planets of the solar system, in the absence of frictional forces. The energy of such a system is constant and the system is said to be *conservative*. The motion of a system of two bodies can be solved exactly, and that of three or more bodies by methods of successive approximation. Classical mechanics has been well established since the seventeenth century.

Thermal physics is concerned with the behaviour of systems containing *large* numbers of particles and, not surprisingly, developed much later than classical mechanics. The full understanding of the microscopic (atomic) approach to thermal physics—called statistical mechanics—was not possible before the development of the quantum theory, but the alternative approach in terms of macroscopic quantities such as pressure and temperature was one of the great triumphs of nineteenth-century physics, and remains of the greatest importance today.

As an example of the methods of classical mechanics consider the simple case of a single particle with mass m_0 in a gravitational field in the z-direction. The change in the *potential energy* of the particle between two points z_A and z_B is

$$U(z_B) - U(z_A) = m_0 g(z_B - z_A) \quad (z_B > z_A) \tag{1.1}$$

where g is the acceleration due to gravity. In a conservative system this increase in the potential energy of the particle is equal to the *work* done on the particle

$$W = \int_{z_A}^{z_B} m_0 g \, dz = m_0 g (z_B - z_A). \tag{1.2}$$

There are two important points to notice about eqn 1.1. Firstly it is only the *external* aspect of the particle which is under consideration not its atomic structure. Secondly, the change in the potential energy depends only on the coordinates of the initial and final points and not on the route (*path*) taken between the two points. The variable z may be said to define the *state* of the *system* and the potential energy is called a *function of state*. An important feature of a function of state is that it is unchanged after a complete *cycle*, that is, on returning to z_A from z_B. Hence

$$\Delta U = 0 \text{ (cycle)}. \tag{1.3}$$

When a body possesses both kinetic and potential energy, the total energy is in general a function of six variables

$$E = \tfrac{1}{2} m_0 (v_x^2 + v_y^2 + v_z^2) + U(x, y, z) \tag{1.4}$$

where v_x is the velocity in the x-direction. The behaviour of the body as a function of time may be calculated if the potential energy and the initial coordinates of position and velocity of the body are known.

Now a cubic centimetre of a solid contains some 10^{22} atoms. A complete description of the behaviour of the atoms in this sample of the solid (the *system of interest* or simply the *system*) would require an equation of the form of eqn 1.4 where the energy is now a function of 6×10^{22} variables and the specification of the initial conditions requires 3×10^{22} coordinates of velocity and an equal number of position. It is clear that an attempt to solve this equation directly by the methods of classical mechanics is unlikely to be successful, although with the aid of fast computers it is now possible to perform calculations for a gas of some two hundred particles, as is discussed in the next section. A further complication, however, is that the motion of atomic particles must be treated by quantum, not classical, mechanics and then, because of the Heisenberg uncertainty principle, it is not possible simultaneously to specify the exact coordinates of position and velocity of each particle.

The general case of a system containing a large number of particles is just as complicated as it appears at first, and each situation has to be treated separately. It is a fact of common observation, however, that if a system is in some way 'isolated' from its surroundings and left undisturbed for a sufficient period of time (just what is meant by this is discussed in Section 1.2), the properties of the system become independent of time. The system is then said to be in *thermal equilibrium*.

The reason for the equilibrium state being of particular importance can be seen by reconsidering the case of the solid sample containing some 10^{22} atoms. The exact behaviour of an *individual* atom still cannot be calculated but the *average* properties of the *whole* system are now independent of time and it will be seen later that it is these average properties which are really of interest. The appearance of the word average makes it clear that we are concerned with the statistical properties of the system, and the microscopic study of systems in thermal equilibrium is called *statistical mechanics*. The earliest example of this approach in physics was the kinetic theory of gases, which is discussed later in this section.

An alternative approach to a system in thermal equilibrium is to ignore the internal (atomic) structure of the system and to observe that it may be completely specified in terms of a small number of *macroscopic* quantities such as pressure, volume, and temperature. This is the approach of *thermodynamics*, one of the objectives of which is to find relationships between these measurable quantities which will apply to all systems in thermal equilibrium.

The mechanical energy of a body, eqn 1.4, is concerned only with its external aspects. Thermal physics is concerned with the *internal energy* of a body and therefore with its relative 'hotness'. In Chapter 2 it will be shown that a number (the temperature) may be associated with the relative 'hotness' of a body at or near thermal equilibrium such that the number increases with the 'hotness'. We shall consider the concept of thermal equilibrium in a little more detail before making the concept of temperature quantitative.

1.2 Thermal equilibrium

When the properties of a thermally isolated system become independent of time the system is said to be in thermal equilibrium. Consider the experimental system and thermal reservoirs of Fig. 1.1. Initially the reservoirs are at temperatures T_{R1} and T_{R2}. The system and reservoir will now exchange heat—a complicated time-dependent process—until finally they come to a common temperature and remain at that temperature. The *approach* to equilibrium would depend on the nature of the experimental system but the properties of all systems in thermal equilibrium are contained in the laws of thermodynamics. The simplicity and the universal character of the thermal equilibrium state makes it of great theoretical importance although it is clear that only carefully prepared experiments are in general in thermal equilibrium with their surroundings.

The equilibrium state should be distinguished from the 'steady state' often considered in the study of transport properties such as thermal conductivity. In Fig. 1.1 the thermal reservoirs are at temperatures T_{R1} and T_{R2} and the temperature along the metal rod will quickly seem to become independent of time. The rod is now in a steady state—the simplest type of transport

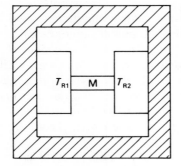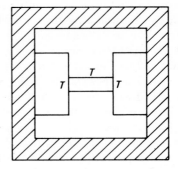

Fig. 1.1 The thermal reservoirs are initially at temperature T_{R1} and T_{R2}. The temperature along the metal rod will quickly seem to become independent of time, varying from T_{R1} at one end to T_{R2} at the other. The rod is said to be in a *steady state* and the heat flow along the rod is governed by its thermal conductivity. Finally the whole system of rod and reservoirs must come to a common temperature. Only then is the system in thermal equilibrium.

situation—but the only possible equilibrium state for the *isolated* system of reservoirs and rod is when T_{R1} is equal to T_{R2}.

This book is, apart from Chapter 9, entirely devoted to systems in thermal equilibrium and it will simply be assumed that such an equilibrium state is in fact attainable. One example will be given below, however, of the passage of a particularly simple system from an arbitrary initial state to a final equilibrium state. The time required for such a transition is called a *relaxation time*. The relaxation times for different systems range from small fractions of a second to times too long to be measured in normal experiments. Systems with very long relaxation times are said to be in *metastable states* and, from the macroscopic point of view, the metastable state can often be treated as if it were a true equilibrium state.

The simplest system for which the approach to equilibrium can be treated in detail is that of a gas of hard spheres. The reasons why real gas molecules can sometimes be treated as classical hard elastic spheres, rather than as scattering potentials in a wave mechanical calculation, are discussed in Section 9.1. The full equations of motion of a set of a hundred such spheres can be solved using a high-speed computer and the evolution from an arbitrary initial state to equilibrium followed in detail.

Suppose, for example, that initially all the spheres have been given the same speed v and therefore kinetic energy $\frac{1}{2}mv^2$. The spheres exchange energy by making elastic collisions with each other and finally reach the equilibrium distribution of energies which is independent of time. The energy distribution is found to approximate to the equilibrium (Maxwell) distribution after about

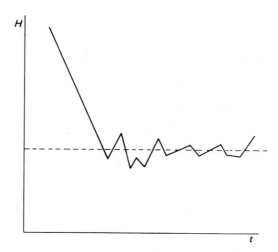

Fig. 1.2 A gas of spheres all of which initially have the same speed are computed to come to the thermal equilibrium distribution (dotted line) after about two collisions per sphere. The gas then performs small random fluctuations about the equilibrium value. The function H is defined in Chapter 6.

two collisions per molecule (Fig. 1.2). In a gas at NTP this time is $\approx 10^{-9}$ s, much less than the time taken to perform a change of volume in a normal laboratory experiment, but other non-equilibrium states such as a variation in density through the gas would take much longer to decay. Notice in Fig. 1.2 that fluctuations continue to exist after the system has reached thermal equilibrium.

When the relaxation times of different parts of a system are very different it is sometimes possible to observe thermal equilibrium within each part separately. The nuclei of the atoms in an insulating solid, for example, may be brought to any common temperature (T_s) without affecting the temperature of the rest of the solid (T_l) (Section 7.8) because the inter-nuclear relaxation time is much shorter than the nuclear–lattice relaxation time (τ_1). When the solid is thermally isolated the nuclear temperature will change towards T_l at a rate determined by τ_1 until a common temperature is reached.

1.3 An outline of the following chapters

Chapters 2–5 are concerned with the macroscopic view of the thermal equilibrium state. The laws of thermodynamics and the concept of a function of state are discussed and applied to some important systems. In Chapters 6 and 7 a microscopic view is adopted and the connection between the two viewpoints is also established. Chapter 10 contains no new principles but seeks to illustrate the power of the results obtained in earlier chapters by applying thermodynamics and statistical mechanics to areas of physics of current research interest. Chapters 8 and 9 are devoted to the kinetic theory of the perfect gas and to the nonequilibrium (transport) properties of systems.

PART I
Thermodynamics

2

First law of thermodynamics

Classical thermodynamics is concerned only with systems in thermal equilibrium. A thermodynamic approach to transport properties such as thermal conductivity and to biological processes is an important branch of current research, but will not be considered in this book.

When a macroscopic system such as a fixed volume of a gas is isolated from its surroundings, its bulk parameters become independent of time after some characteristic time for the system has elapsed. A system will be said to be in thermal equilibrium if its macroscopic state can be *completely* specified by some small number of macroscopic parameters which are all independent of time. At the microscopic (atomic) level processes such as collisions of gas molecules continue to lead to changes in the states of individual molecules but macroscopic quantities will be seen to involve averages over all the molecules, and these average values become independent of time once the system has reached thermal equilibrium.

A *simple system* will be defined as a system which in thermal equilibrium can be completely specified by just two independent variables. A volume of gas of fixed mass in which no chemical reactions occur is an example of a simple system, since the state of the gas is completely specified by the pressure and volume. A volume of water near 4°C is a non-simple system (Fig. 2.1), since three independent parameters (P, V, T) are needed to completely specify the equilibrium state.

2.1 Zeroth law and scale of temperature

The most directly accessible thermal concept is not heat (which will be discussed in Section 2.3) but rather temperature, the relative sensations of hot and cold. When a hot body (A) is placed in physical contact with a cold one (B) the composite body is always found to have a 'hotness' intermediate between that of A and B "once thermal equilibrium has been reached" Reseparating the bodies does not lead to them returning to their original states of hotness and coldness. A body which is hot relative to its surroundings cools down. There is therefore a definite *direction* in time for thermal processes which is absent from purely mechanical interactions.

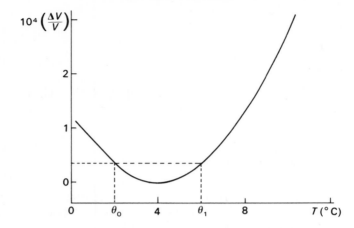

Fig. 2.1   The temperature dependence of the volume of unit mass of water at constant pressure. A specification of (P, V_0) near $4\,°C$ is not sufficient to define the state of the system since the water temperature could be θ_0 or θ_1.

The argument may be made more concrete by considering two containers of gas of equal mass and volume which are completely isolated from external influences (Fig. 2.2). The walls of such a container are said to be adiabatic walls. The gases are each in thermal equilibrium within their own containers and the original states of the two gases are completely defined by the parameters (P_{A1}, V) and (P_{A2}, V). If P_{A1} is greater than P_{A2}, then when the gases are placed so that the two containers are in thermal contact, say by using a thin metal (diathermal) wall then P_{A1} will decrease to P_{B1} and P_{A2} will increase to P_{B2}. When the two containers are reseparated within adiabatic walls these new states will continue to exist. The thermal process is therefore *irreversible*. The second law of thermodynamics discussed in Chapter 3 is the great generalization of the difference between mechanical and thermal processes. The zeroth law, which is discussed in this section is not necessary for a completely logical development of thermodynamics, but remains a useful introduction to thermal concepts.

All the possible equilibrium states of the simple system labelled 1 may be described by some function of two variables $f(P_1, V_1)$ since the parameters (P_1, V_1) completely define the state. Similarly system 2 will have states given by $f(P_2, V_2)$. When these two systems, each in thermal equilibrium, are placed in thermal contact, however, the parameters of both systems will in general change with time until a new state of thermal equilibrium of the whole system has been established. In the special case where the parameters

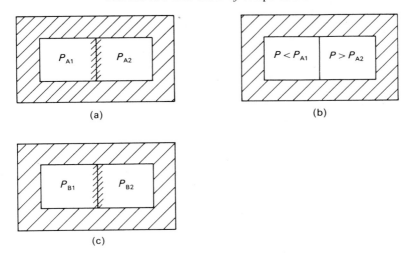

Fig. 2.2 (a) The two systems enclosed within adiabatic walls are each in internal thermal equilibrium. (b) When thermal contact is allowed however the pressure in each container changes until finally the composite system comes to a new equilibrium state. (c) If the adiabatic wall is now replaced the individual systems remain in these states.

do not change when the systems are placed in thermal contact the systems 1 and 2 are said to be in mutual thermal equilibrium.

First consider system 1 in the state (P_{A1}, V_{A1}) as the reference system. Then $f(P_{A1}, V_{A1})$ is equal to some value θ_{A1} and all the states of system 2 which are in thermal equilibrium with system 1 are classified by the equation

$$f(P_{A1}, V_{A1}) = \theta_{A1} = f(P_2, V_2) \tag{2.1}$$

The function $f(P_2, V_2)$ is called an isotherm (Fig. 2.3). Now using system 2 as the reference system all the states of system 1 which satisfy eqn 2.1 can be classified so that finally

$$f(P_1, V_1) = f(P_2, V_2) = \theta \tag{2.2}$$

classifies all possible states of mutual equilibrium of the two systems.

The *separate* equilibrium states of the two systems involve four independent variables but the *mutual* equilibrium states involve only three independent variables. The restraint imposed by eqn 2.2 is called the empirical temperature and an equation of the form

$$f(P, V) = \theta \tag{2.3}$$

is called an equation of state.

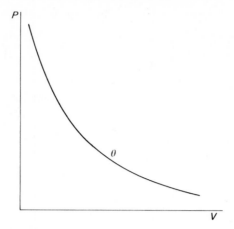

Fig. 2.3 The locus of all the states of a system which are in mutual thermal equilibrium with a reference system is called an isotherm. θ (defined by eqn 2.2) is called the *empirical* temperature.

The zeroth law may now be stated: if system 1 is in mutual thermal equilibrium with system 2 and with system 3, then system 2 is in mutual thermal equilibrium with system 3.

System 1 can therefore be used to classify all states of mutual equilibrium between other systems, that is to say all the states of a system which lie on the isotherm having the empirical temperature θ, where θ is defined by eqn 2.3. System 1 used in this way is a thermometer.

A *scale* of temperature may be defined in terms of eqn 2.3 if a value of θ is decided upon for some easily reproduceable fixed point or points. The Celsius scale is defined, for example, if some property of system 1 is measured, say the pressure of a gas at constant volume, in melting ice (0°C) and the boiling point of water under a pressure of one atmosphere (100°C). Then any other temperature on the Celsius scale of system 1 is *defined* by the equation

$$\theta_P = \frac{P_\theta - P_0}{P_{100} - P_0} \times 100°C \tag{2.4}$$

The values at the measured points are therefore simply joined by a straight line (Fig. 2.4) and extrapolated as necessary.

The Celsius scale is still in everyday use but clearly has no fundamental significance. In particular it should be noticed that there is no restriction on the *range* of temperature since eqn 2.4 may be extrapolated to plus and minus infinity.

The temperature of a given simple system is uniquely defined by eqn 2.4, and all other systems in mutual thermal equilibrium with the reference system are correctly classified but the *number* relating to the temperature will depend

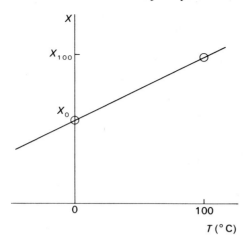

Fig. 2.4 The Celsius scale for a given thermometric substance is established by measuring the magnitude of some parameter at the melting point of ice and the boiling point of water under a pressure of 1 atmosphere. All other temperatures are found by measuring the parameter and either reading off the temperature from the graph or by using an equation of the form of eqn 2.4.

upon the thermometer. Consider two thermometers, say the length of a column of liquid such as mercury and the resistance of a metal. Then

$$\theta_l = \frac{l_\theta - l_0}{l_{100} - l_0} \times 100°C$$

$$\theta_R = \frac{R_\theta - R_0}{R_{100} - R_0} \times 100°C$$

by definition.

Suppose however that the resistance of the wire is measured as a function of temperature (using the mercury thermometer) and the resistance is found to have the form

$$R_\theta = R_0(1 + b\theta_l + c\theta_l^2)$$

where R_0, b, c are constants. Then

$$\theta_R = \frac{b\theta_l + c\theta_l^2}{b + 100c} \tag{2.5}$$

and θ_R is not equal to θ_l (except at the fixed points) unless c is zero (Fig. 2.5). It is therefore not sufficient when temperature is defined by eqn 2.4 to say that a solid melts at 50°C, the particular thermometer must also be specified. The Celsius scale as defined above is therefore unsatisfactory because (a) no

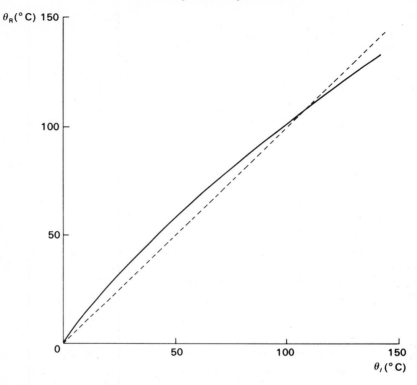

Fig. 2.5 The relationship between the temperature measured using a mercury thermo-meter and a resistance wire (eqn 2.5). The constant c has been taken to be negative as in Exercise 2.1.

physical reasoning lies behind the scale, and (b) the number given for the temperature of a body will depend upon the thermometer in use. The 'fixed points' of the Celsius scale are also unsatisfactory because they cannot be reproduced experimentally to the highest accuracy required in modern thermometry ($\approx 0.001\,°C$).

The objection to the Celsius scale made in point (a) above can only be fully understood after the second law of thermodynamics has been established (in Section 3.1). The objection in point (b) could be met by always using the same thermometer. The perfect gas scale of temperature is the chosen scale because, as shown in Section 3.1, it is identical to the thermodynamic scale of temperature, which is independent of the thermometric substance.

The perfect gas scale is defined using the equation of state

$$T = \frac{PV}{R}$$

for 1 mole of a perfect gas, where R is the gas constant. (The concept of a perfect gas arose from the experimental observation that the equation of state of real gases, Section 2.2, tended to eqn 2.6 at sufficiently high temperature and low density).

The size of the degree is defined by the *triple point* of water, where ice, water and water vapour are in equilibrium ($\approx 0.01°C$) being taken as 273.16 K where K means degrees Kelvin (after the inventor of the thermodynamic scale). Then any other temperature is defined by

$$T = \frac{(PV)_T}{(PV)_{273.16}} \times 273.16 \text{ K}. \tag{2.7}$$

The thermodynamic scale may be shown to be always positive (once the triple point has been defined to be positive) and therefore implies an absolute zero of temperature (Section 3.1).

The Celsius scale is now treated as a derived quantity defined by

$$\theta = (T_K - 273.15)°C.$$

The choice of 273.16 K for the triple point was made so that the fixed points on the Celsius scale were nearly unaltered from their original values.

The actual measurement of temperature on the perfect gas scale is extremely difficult. A real gas will not have an equation of state given by eqn 2.6 but for sufficiently high temperatures an extrapolation to zero density of the product (PV) does lead to the gas scale (Section 2.2). Then

$$T = \left[\frac{(PV)_T}{(PV)_{273.16}} \right]_{P \to 0} \times 273.16 \text{ K}.$$

A direct measurement of temperature on the thermodynamic scale is however rarely attempted outside a standards laboratory such as the National Physical Laboratory (UK) or the National Bureau of Standards (USA).

The International Practical Temperature Scale (IPTS) is a set of reproduceable fixed points, some of whose values on the thermodynamic scale are given in Table 2.1. A convenient thermometer, the resistance of platinum wire in the temperature range 14–900 K for example, is then chosen and its properties measured at the fixed points within its range. An interpolation between the fixed points, either by curve-fitting using a computer or else by a correction formula similar to eqn 2.5, enables a table of resistance against thermodynamic temperature to be constructed. The scale is of course not *linear* in the resistance since the resistance of platinum is never exactly proportional to the thermodynamic temperatures. In Table 2.2 some values of

$$z = \frac{T}{R} \left(\frac{dR}{dT} \right) = \left(\frac{d \ln R}{d \ln T} \right) \tag{2.8}$$

Table 2.1. The International Practical
Temperature Scale

Fixed point	(K)
Triple point of hydrogen	13.81
Boiling point of hydrogen	20.28
Boiling point of oxygen	90.19
Triple point of water	273.16
Boiling point of water	373.15
Freezing point of zinc	692.73
Freezing point of silver	1235.08
Freezing point of gold	1337.58

Some of the important reference temperatures on the IPTS. The pressure is understood to be exactly 1 atmosphere for the boiling and freezing points.

Table 2.2. The performance of two typical resistance thermometers as a function of temperature

Temperature	Platinum thermometer		Germanium thermometer	
$T(K)$	$R(\Omega)$	$\dfrac{T}{R}\left(\dfrac{dR}{dT}\right)$	$R(\Omega)$	$\dfrac{T}{R}\left(\dfrac{dR}{dT}\right)$
1.5	—	—	13449	3.74
4.2	0.0113	1.48	1099	1.62
20	0.1073	3.52	193.36	3.89
100	7.3387	1.50	4.868	0.45
300	28.318	1.07	—	—
770	72.536	0.92	—	—

The performance of a platinum resistance thermometer is seen to be inferior to that of a typical germanium thermometer at temperatures below 20 K but becomes greatly superior to it at high temperature.

are given for platinum and germanium resistance thermometers. The value of z is a rough measure of the sensitivity of the thermometer and would be equal to 1 if the resistance was directly proportional to the thermodynamic temperature. The germanium resistance thermometer can be seen to have low sensitivity at high temperature—where the platinum resistance thermometer is obviously superior—but improves as the temperature is lowered. The measurement of temperatures on the thermodynamic scale at low temperature presents extra difficulties, one solution to which is discussed in Section 10.6.

2.2 Equation of state

A simple system has been defined as one whose state is completely specified by two independent variables. The equation which contains all equilibrium states of the system is therefore of the form

$$T = f(P, V)$$

or equivalently

$$P = f(V, T)$$
$$V = f(P, T).$$

The two parameters need not be pressure and volume, which have been used up to now because a gas has been used for illustrative examples. For the stretching of a wire at constant volume, for example, they would be tension and length. The important feature of a simple system is that T is a function of only two variables so that

$$T = f(x, y) = T(x, y).$$

The second expression will be used from now on to indicate that T is some function of x and y.

The simplest equation of state is that of the perfect gas which may be written in a number of equivalent forms, e.g.

$$PV = n_0 RT = m_0 R_m T = n_0 N_A kT = NkT = nVkT \qquad (2.9)$$
$$R = MR_m = N_A k$$

for n_0 moles of gas of total mass m_0 and molecular weight M. The values of the gas constant per mole (R), Avogadro's number (N_A) and the Boltzmann constant (k) are given in Appendix V and the molecular weights of some gases in Appendix VI.

The isotherms of a perfect gas on a P–V diagram (Fig. 2.6) are hyperbolae. At low temperature and high density however, a real gas will liquefy and one section of each isotherm is parallel to the V-axis (Fig. 2.6). Equation 2.9 is derived in Section 7.5 for a gas of particles of negligible total volume whose only interactions are elastic collisions. The energy of the particles is therefore purely kinetic energy. At low temperature and high density these restrictions are no longer valid for a real gas and a more complicated equation of state is required. The virial expansion is one such equation of state and may be written either

$$PV = A\left(1 + \frac{B_V}{V} + \frac{C_V}{V^2} + \cdots\right) \qquad (2.10)$$

or

$$PV = A + B_P P + C_P P^2 + \cdots. \qquad (2.11)$$

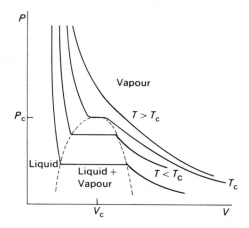

Fig. 2.6 The isotherms of a real fluid. At high temperature and low density (large volume) the perfect gas eqn (2.9) is satisfactory but at temperatures below T_c the gas liquefies. The perfect gas equation is therefore unsatisfactory except for $T \gg T_c$ and low density.

The series may be understood as the corrections to the perfect gas law due to pair interactions between molecules (B), three-body interactions (C), and so on. The coefficients A, B, and C are functions of temperature. The coefficient A is simply equal to RT as shown in Exercise 2.2. The virial expansion is most useful⁻ for relatively small departures from the ideal gas law ($C_V V^{-2} \ll B_V V^{-1} \ll 1$) where simply B_V is often a sufficient correction. The series does not converge as the temperature approaches the critical temperature (T_c in Fig. 2.6).

The virial coefficients can be measured directly for real gases and also calculated for certain models of intermolecular interactions. A more ambitious approach is to attempt to find an equation containing parameters *independent* of the temperature which would reduce to the perfect gas law in the limit of high temperature and low density, but would also fit the other isotherms in Fig. 2.6. The first attempt at this type of equation of state was due to van der Waals, who introduced two constants, 'a' to correct for intermolecular attraction, and 'b' to correct for the finite size of the molecules

$$\left(P + \frac{a}{V^2}\right)\left(V - b\right) = RT \tag{2.12}$$

for mole 1. The value of b is similar for all gases but a varies greatly e.g He(O_2): $b = 2.3(3.2) \times 10^{-5}$ m³ mol⁻¹, $a = 3.4(140) \times 10^{-3}$ J m³ mol⁻².

The isotherms of this equation are shown in Fig. 2.7. The equation is cubic in V and therefore has one or three real roots. The isotherms are clearly not identical to Fig. 2.6. The states along AB and CD are metastable and may be

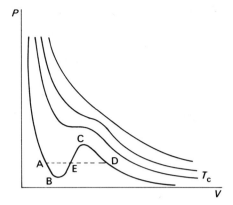

Fig. 2.7 Isotherms of a van der Waals gas. The states along AB, CD are measurable. BC is unstable. The dotted line shows the Maxwell construction.

reached experimentally. The portion BC is, however, mechanically unstable since $(\partial P/\partial V)_T$ is positive. (A small decrease in the pressure in region BC would lead to a decrease in the volume, which is clearly unstable). Maxwell has shown that the best van der Waals isotherm is that for which the area ABE is equal to ECD (see Fig. 2.7).

Many other two-parameter functions of state have been introduced since the van der Waals equation but none of them are able to reproduce all the features of a real fluid. The van der Waals equation therefore remains a useful approximation and will be used throughout the book to distinguish between the properties of real and perfect gases.

The flat section of the isotherms of a real fluid shown in Fig. 2.6 can be seen to become smaller as the temperature rises. At the critical temperature T_c it becomes a point of inflexion (P_c, V_c) and therefore

$$\left(\frac{\partial^2 P}{\partial V^2}\right)_T = \left(\frac{\partial P}{\partial V}\right)_T = 0 \tag{2.13}$$

at the critical point. Using eqn 2.12

$$P_c = \frac{a}{27b^2} \quad V_c = 3b \quad T_c = \frac{8a}{27Rb} \tag{2.14}$$

$$\frac{RT_c}{P_c V_c} = \frac{8}{3} \tag{2.15}$$

$$\left(P' + \frac{3}{(V')^2}\right)\left(3V' - 1\right) = 8T'$$

$$P' = \frac{P}{P_c} \quad V' = \frac{V}{V_c} \quad T' = \frac{T}{T_c}. \tag{2.16}$$

The van der Waals equation therefore predicts a value of 8/3 for $RT_c/P_c V_c$ for *all* gases rather than the perfect gas value of unity. The experimental values are however always greater than three and also depend upon the gas. For example, $RT_c/P_c V_c$ for hydrogen is 3.06 and for helium is 3.08 but for nitrogen and oxygen it is 3.42. The reduced equation of state given in eqn 2.16 is a universal curve which all gases would obey if the van der Waals equation was correct. In fact any two-parameter equation of state may be brought into reduced form.

The three equations of state introduced in this section will be used throughout the book to observe the difference in the behaviour of real and perfect gases. The reader must remember, however, that close to the critical point all three equations are inadequate. The gas–liquid transition cannot be followed by equations of this form. A more subtle point to note is that a real gas with an equation of state other than $PV = RT$ may, under cetain special conditions, behave like a perfect gas. In Exercise 2.2 it is shown that $B_P = B_V = b - (a/RT)$. There is therefore a temperature (the Boyle temperature T_B) equal to a/Rb where B is zero. The equation of state given by eqn 2.10 is equivalent to the perfect gas equation $PV = RT$ over a wide range of pressure at the Boyle temperature. However if the property being studied involves not B but, say, dB/dT, the real gas will not behave like a perfect gas at the Boyle temperature. In the throttling process discussed in Section 5.3, for example the important parameter is $[T(dB/dT) - B]$ and the real gas behaves like a perfect gas not at T_B but at $2T_B$.

2.3 First law of thermodynamics

In mechanics, the potential energy of a mass m_0 at a point z where the acceleration due to gravity is g, is defined to be

$$U(z) = m_0 gz.$$

Since the reference plane has been chosen arbitrarily, $U(z)$ is defined only to within a constant. The *difference* of potential energy between two points is, however, completely defined since

$$U(z_B) - U(z_A) = m_0 g(z_B - z_A). \tag{2.17}$$

In a system without dissipative forces such as friction (a conservative system) the increase in the potential energy between z_B and z_A is simply equal to the work done on the mass m

$$W = \int_{z_A}^{z_B} m_0 g \, dz = m_0 g(z_B - z_A) = U(z_B) - U(z_A). \tag{2.18}$$

The change in the energy of the system in going from z_B to z_A depends only on the final coordinates and not on the path by which the coordinates were reached (Fig. 2.8).

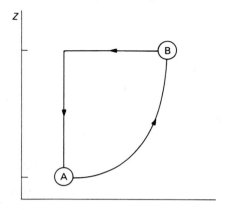

Fig. 2.8 The difference in the potential energy of a particle at two points is independent of the path joining the points. In a conservative system the work done *on* the body to take it from A to B is equal to the work done *by* the body if it returns from B to A.

The potential energy is just one of many forms of energy. The law of conservation of energy states that energy can neither be created nor destroyed. In a closed system without dissipative forces, the total energy may be changed, say, from potential energy to kinetic energy and back to potential energy, provided that at all times the sum of potential and kinetic energies remains constant.

$$\tfrac{1}{2}m_0 v_B^2 + m_0 g z_B = \tfrac{1}{2}m_0 v_A^2 + m_0 g z_A.$$

It sometimes becomes necessary to extend the definition of energy to ensure that the energy of a closed system is conserved. The change from Newtonian mechanics to relativistic mechanics for example, involved the recognition that the rest mass of a particle m_0 contains energy $m_0 c^2$ where c is the velocity of light. A decrease of mass in a closed system therefore involves the release of energy (as occurs in a nuclear reaction). When all sources of energy have been taken into account however, the law of conservation of energy appears to be completely rigorous.

There are three features of eqn 2.18 which should be noted:

(1) the equation may be read in either direction. If the particle is at z_A then work (W) may be done *on* the mass to move it to z_B. Alternatively if the particle is at z_B it is available to *do* work W as it returns to z_A;

(2) the temperature of the body does not appear in the equation. Mechanics is concerned only with the external aspect of the body;

(3) the change in the potential energy depends only upon the initial and final coordinates, not upon the path by which the coordinates were reached.

In particular if the particle is at z_A and completes a cycle back to z_A the change in the potential energy is zero. This is written

$$\Delta U = \oint mg\, dz = 0. \tag{2.19}$$

The function $U(z)$ is called a function of state since it is defined to within a constant by the coordinates (the state) of the system.

The first law of thermodynamics is an extension of the concept of conservation of energy to include thermal processes. An *internal* energy function U, which for a simple system will be a function of two independent variables, is postulated. $U(T, V)$ for example will be defined only to within an arbitrary constant but $U(T_B, V_B) - U(T_A, V_A)$ is, like the change of potential energy in eqn 2.17, completely defined. We will simplify the notation to $U(B)$ where it is to be understood that B defines the state of the system.

The question remains of how to establish the properties of the internal energy function and indeed to confirm its existence by experiment. The clue is provided by eqn 2.18. A system in thermal equilibrium in state A is confined within rigid adiabatic walls. The system is therefore isolated from all outside influences but it is still possible to perform work on the system (Fig. 2.9). The experiments of Joule showed that regardless of whether the work performed under adiabatic conditions was mechanical or electrical, a quantity of work W always led to the same final state B of the system.

$$W_{ad} = U(B) - U(A) \tag{2.20}$$

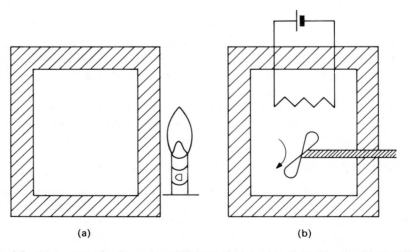

(a)   (b)

Fig. 2.9   (a) A system in thermal equilibrium within rigid adiabatic walls. The system is isolated from outside influences. (b) The state of the system may be changed by doing work on the system by either mechanical or other means.

The internal energy function of any system is therefore to be found by performing work on the system under adiabatic conditions. (The meaning of this equation will become clearer in eqn 2.51.)

When the system is in state A but not surrounded by adiabatic walls, the work W required to take the system to state B is found to depend upon the path by which the process takes place. The difference between W_{ad} and W is called the quantity of heat, or simply the heat, added to the system.

$$U(B) - U(A) = W + Q. \tag{2.21}$$

The sign convention here is that W is positive if work is done *on* the system and Q is positive if heat is added *to* the system. In older text books the convention W is positive if work is done *by* the system, is used and eqn 2.21 would then contain $-W$. The first law of thermodynamics is therefore a law of the conservation of energy which includes quantity of heat as a form of energy transfer.

2.4 The reversible quasistatic process

The function U is to be considered as a continuous differentiable single-valued function of the temperature and of one other coordinate in a simple system. The differential dU is therefore well defined. It is a perfect differential of two variables for a simple system say (T, V) and

$$dU = \left(\frac{\partial U}{\partial T}\right)_V dT + \left(\frac{\partial U}{\partial V}\right)_T dV. \tag{2.22}$$

The differential of a function of two or more variables is discussed in Appendix I. $U(B) - U(A)$ is independent of the path between states A and B and may be obtained by integrating eqn 2.22.

When Q is zero (the adiabatic process), eqn 2.20 is regained and W is a state function. Similarly if W is zero, Q is a state function. In general however, the magnitude of W will depend upon the path, as will be shown in Section 2.5, and since $U(B) - U(A)$ is independent of the path, Q also changes to compensate for changes in W. There are no unique functions of state W and Q and therefore no perfect differentials dW and dQ (see Appendix I).

When the infinitesimal change dU is carried out so slowly that the process takes a long time compared to the longest relaxation time of the system, the change is said to be a quasistatic process. If reversing the small change in the parameters of the system also takes the entire system back to its original state, the process is a reversible quasistatic process. The process must take place not only sufficiently slowly but also without friction. Under these special conditions the infinitesimal work term is well defined and may be written dW_R. The symbol d rather than d is used to emphasise that there is no unique function of state W. Similarly the heat change may be written dQ_R.

The suffix R restricts the changes to reversible processes. Then

$$dU = dQ_R + dW_R \qquad (2.23)$$

is the statement of the first law of thermodynamics for an infinitesimal reversible change of state. The right-hand side of the equation may be integrated from state A to state B if the reversible path between the two states is specified. dQ_R and dW_R are called imperfect differentials.

2.5 Work

Consider the reversible quasistatic change shown in Fig. 2.10. The gas in the cylinder expands slowly against a pressure equal (to within an infinitesimal variation) to its own. Then the work done *on* the gas is given by

$$dW_R = -F\,dx = -PA\,dx = -P\,dV.$$

The total work done in a reversible change from state A to state B is

$$W = -\int_A^B P\,dV \qquad (2.24)$$

The work done is therefore simply the area between the path from A to B and the V-axis (Fig. 2.10). The work done depends upon the path. In a complete cycle the change in the internal energy is zero, but the work done

$$W = -\oint P\,dV \qquad (2.25)$$

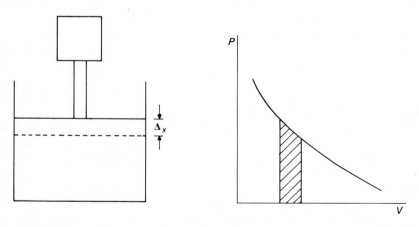

Fig. 2.10 A reversible quasistatic expansion of a gas in a cylinder of cross-sectional area A. The work is the area between the path and the V-axis.

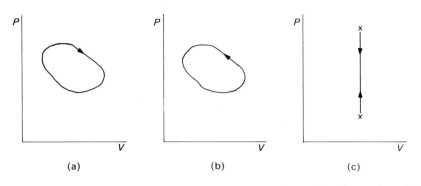

Fig. 2.11 In a cycle the change in the internal energy is zero but the total work is equal to the area enclosed by the cycle. (a) Net work done by the gas, (b) net work done on the gas, (c) net work zero.

may be positive, negative or zero (Fig. 2.11). Notice that

$$\oint_{A} P\,dV = -\oint_{A} P\,dV$$

where the arrow indicates that the cycle is performed clockwise or anticlockwise from A along the same reversible path in each case.

The calculation of W using eqn 2.24 therefore requires the equation of state of the system to relate P to V, and also the path by which the system moves from A to B. Two simple examples will now be given. If the system is one mole of a perfect gas, then $PV = RT$ and

$$W = -R \int_{V_A}^{V_B} \left(\frac{T}{V}\right) dV.$$

In a reversible quasistatic change from V_A to V_B at constant temperature T_0 (an isothermal change) the integral becomes

$$W = -RT_0 \int_{V_A}^{V_B} \frac{dV}{V} = -RT_0 \ln\left(\frac{V_B}{V_A}\right). \qquad (2.26)$$

An even simpler example is a change at constant pressure P_0. In this case the equation of state is not required since

$$W = -P_0 \int_{V_A}^{V_B} dV = -P_0(V_B - V_A). \qquad (2.27)$$

The work term is not necessarily of the form $P\,dV$, but in a simple system with only two independent parameters can always be written in the form $X\,dx$. A wire extended dl under a tension $\mathscr{F}$ at constant volume has a work term $\mathscr{F}\,dl$, for example. This term is positive because l increases with $\mathscr{F}$, unlike V which decreases as P increases. The work term for a system in a magnetic field is discussed in Section 10.5

2.6 Heat capacity

When a quantity of heat dQ_R is added to a system the temperature of the body usually increases. The heat capacity of the body is defined as

$$C_x = \left(\frac{dQ_R}{dT} \right)_x .$$

(2.28)

Here x refers to the constraint under which C is to be measured. Since Q is not a function of state the heat capacity is meaningless unless the constraint is specified. The most important quantities are the heat capacity at constant pressure

$$C_P = \left(\frac{dQ_R}{dT} \right)_P$$

(2.29)

because this is the quantity usually measured for a solid, and the heat capacity at constant volume

$$C_V = \left(\frac{dQ_R}{dT} \right)_V$$

(2.30)

which is the quantity calculated by statistical mechanics. The heat capacity under conditions of saturated vapour pressure is also important for liquids.

The specific heat capacity is the heat capacity of unit mass of a substance. The unit is usually taken to be the mole in thermodynamics and the molecule in statistical mechanics. The relation between the two is then simply the Avagadro constant.

When a system changes its phase (solid–liquid, liquid–vapour, solid–vapour) the heat absorbed per unit mass by, say, the solid at the melting point is called the specific latent heat. The specific heat capacity at the phase transition is infinite, since the system absorbs a quantity of heat without change of temperature.

The specific heat capacity of a substance is in general a function of both temperature and volume and as the most experimentally accessible thermal quantity is of great theoretical and experimental importance. The specific heat capacity of solids is discussed in Section 10.3 and of gases in Section 7.5.

A relationship between C_P and C_V of any system may be found using only the first law of thermodynamics, but it will be possible to simplify the general result after the second law has been established. The first law for an infinitesimal, reversible quasistatic change of a simple system may be written using eqns 2.22–2.24.

$$\mathrm{d}Q_R = \left(\frac{\partial U}{\partial T}\right)_V \mathrm{d}T + \left[P + \left(\frac{\partial U}{\partial V}\right)_T\right]\mathrm{d}V. \tag{2.31}$$

Therefore

$$\left(\frac{\mathrm{d}Q_R}{\mathrm{d}T}\right)_V = \left(\frac{\partial U}{\partial T}\right)_V = C_V \tag{2.32}$$

giving

$$\mathrm{d}Q_R = C_V\,\mathrm{d}T + \left[P + \left(\frac{\partial U}{\partial V}\right)_T\right]\mathrm{d}V \tag{2.33}$$

and

$$\left(\frac{\mathrm{d}Q_R}{\mathrm{d}T}\right)_P = C_V + \left[P + \left(\frac{\partial U}{\partial V}\right)_T\right]\left(\frac{\partial V}{\partial T}\right)_P.$$

The left-hand side of the last equation is equal to C_P so

$$C_P - C_V = \left[P + \left(\frac{\partial U}{\partial V}\right)_T\right]\left(\frac{\partial V}{\partial T}\right)_P \tag{2.34}$$

is the general result. The term $P(\partial V/\partial T)_P$ may be evaluated from the equation of state of a particular system, but further information is required to find $(\partial U/\partial V)_T$. This term can, in fact, be eliminated after the second law of thermodynamics has been established in Section 3.1. The internal energy of a perfect gas, however, is independent of its volume (Joule's law) so $(\partial U/\partial V)_T$ is zero. Then

$$P\left(\frac{\partial V}{\partial T}\right)_P = P\frac{\partial}{\partial T}\left(\frac{RT}{P}\right)_P = R$$

for one mole of a perfect gas and

$$C_P - C_V = R = N_A k. \tag{2.35}$$

Notice that the specific heat capacities of a perfect gas with two or more atoms per molecule may be a function of temperature. At any given temperature however the *difference* in the molar specific heat capacities is equal to R. The difference in the heat capacities per molecule of a perfect gas may be written

$$c_P - c_V = k \tag{2.36}$$

where k is the Boltzmann constant.

2.7 Heat engines

Thermodynamics began as the study of the efficiency of heat engines. An engine will be defined as a machine which at the end of a complete cycle has converted some heat into useful work. The process can then be repeated for another cycle and so on. Since ΔU is zero for a complete cycle the first law (eqn 2.21) becomes simply

$$W = -Q. \tag{2.37}$$

The simplest possible engine is represented in Fig. 2.12. The engine extracts heat of magnitude Q_A from the hot body, performs work W, rejects heat Q_B to the cold body and returns to its original state. When the thermal capacity of the bodies is so large that the change of Q_A or Q_B does not effectively change their temperature they are called thermal reservoirs. The work and heat processes in the cycle are related by eqn 2.37

$$W = Q_B - Q_A$$

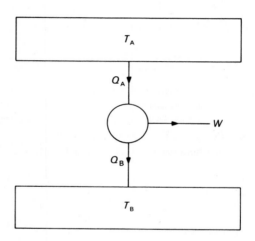

Fig. 2.12   An engine absorbs heat from one thermal reservoir, does work, rejects heat to a second thermal reservoir at a lower temperature and then returns to its original state. The whole operation is called a cycle.

(remembering that work done *by* the system is negative). The efficiency of an engine (η) is defined to be

$$\eta = \frac{\text{Work done}}{\text{Heat extracted from the hot body}} \tag{2.38}$$

$$= -\frac{W}{Q_A} = \frac{Q_A - Q_B}{Q_A}. \tag{2.39}$$

The statement that Fig. 2.11 is a representation of the simplest *possible* engine cannot be maintained on the basis of the first law of thermodynamics alone. The first law may be stated, on the basis of eqns 2.37 and 2.39, as 'The efficiency of an engine working in a cycle cannot exceed unity'.

$$\eta \leqslant 1 \quad \text{(First Law).} \tag{2.40}$$

The efficiency is unity if no heat is rejected to the cold reservoir ($Q_B = 0$). The second law, to be considered in Chapter 3, imposes the more rigorous condition

$$\eta < 1 \quad \text{(Second Law)} \tag{2.41}$$

that is to say a source *and* a sink of heat are required for an engine since $Q_B \neq 0$. The efficiency of real heat engines is always far less than unity of course, but in Chapter 3 the maximum *theoretical* efficiency of a heat engine will also be shown to be in agreement with eqn 2.41.

There is a limit therefore to the efficiency with which heat can be converted into work by a cyclic process. Heat can however be completely converted to work in any non-cyclic process for which there is no change in the internal energy of the system. A perfect gas, for example, has

$$\left(\frac{\partial U}{\partial T}\right)_V = C_V ; \left(\frac{\partial U}{\partial V}\right)_T = 0.$$

The internal energy of a perfect gas is therefore a function of temperature $U(T)$ only. In a reversible isothermal expansion of a perfect gas $\Delta U = 0$ and

$$Q = - \int dW = RT_0 \ln \left(\frac{V_B}{V_A}\right) \tag{2.42}$$

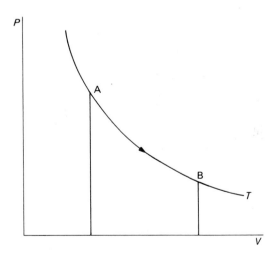

Fig. 2.13 An isothermal expansion of an ideal gas from A to B converts all the heat absorbed from the thermal reservoir to work since the internal energy of a perfect gas depends only upon the temperature, but the process is not cyclic.

from eqn 2.26. Heat Q has been completely converted to work but the system is now in the state B, having extracted heat Q from its surroundings to keep the temperature of the gas constant (Fig. 2.13).

2.8 Conclusions based on the first law

The first law of thermodynamics has been established to be in general

$$U(B) - U(A) = Q + W \tag{2.43}$$

and for a reversible infinitesimal change

$$dU = dQ_R - P\,dV$$
$$dQ_R = dU + P\,dV \tag{2.44}$$

U is a function of state and for a simple system is a function of two independent variables. Considering U as a function of temperature and volume $[U(T, V)]$

$$dQ_R = \left(\frac{\partial U}{\partial T}\right)_V dT + \left[P + \left(\frac{\partial U}{\partial V}\right)_T\right] dV$$

$$dQ_R = C_V\,dT + \left[P + \left(\frac{\partial U}{\partial V}\right)_T\right] dV \tag{2.45}$$

dQ_R is well defined but depends upon the path by which the changes in temperature and volume occur.

The first law in the form of eqn 2.44 for a simple system contains only one work term $(-P\,dV)$. The sum of all the $(n-1)$ reversible work processes occurring in a general system may be written

$$dW_R = \sum_{r=1}^{n-1} X_r\,dx_r$$

where X_r is a generalised force (such as $-P$ or $\mathscr{F}$) and dx_r is a generalised infinitesimal displacement (dV, dl). The internal energy now becomes a function of n variables, say the temperature plus the $(n-1)x_r$ and

$$dU = \left(\frac{\partial U}{\partial T}\right)_x dT + \sum_{r=1}^{n-1} \left(\frac{\partial U}{\partial x_r}\right)_c dx_r .$$

The first law now becomes

$$dQ_R = \left(\frac{\partial U}{\partial T}\right)_x dT + \sum_{r=1}^{n-1} \left[\left(\frac{\partial U}{\partial x_r}\right)_c - X_r\right] dx_r \tag{2.46}$$

$$= \sum_{r=1}^{n} Y_r\,dy_r . \tag{2.47}$$

Here $(\partial U/\partial T)_x$ means that all the $(n-1)$ x_r are to be held constant and $(\partial U/\partial x_r)_c$ means that the temperature and all x other than x_r are to be held constant while performing the differentiation. The final form of eqn 2.47 will be found useful in Chapter 3, the significance of the Y_r and dy_r should be clear from eqn 2.46.

The parameters which occur in eqn 2.44 or 2.46 may be divided into two types. The parameters which are independent of the size of the system (P, $\mathscr{F}$, T) are called *intensive* parameters; those proportional to the size of the system (U, Q, C_V, V, l . . .) are called *extensive* parameters. The internal energy, for example, could be written $U=uV$ where u is the internal energy per unit volume of a uniform system. This division is only correct for large systems where surface effects are unimportant relative to the bulk contribution to the thermodynamic quantities.

In establishing eqn 2.43 in Section 2.3 we began with the purely mechanical equation for the potential energy of a body

$$W = U(z_B) - U(z_A) \tag{2.48}$$

and following Joule's experiments on the conversion of work into heat we wrote, by analogy

$$W_{ad} = U(B) - U(A) \tag{2.49}$$

for a system within adiabatic walls.

Although these two equations appear to be identical there is in fact a most important distinction to be made between them. In Section 2.3 we saw that in mechanics z_B and z_A were on an equal footing (eqn 2.18), either work was done *on* the particle to raise it from z_A to z_B or else work done *by* the particle in going from z_B to z_A. The outstanding result of Joule's experiments for a system within rigid adiabatic walls was that while any *pair* of states (A, B) were found to be accessible by a suitable adiabatic work process, if B could be reached from A then A could *not* be reached from B.

The difference between eqns 2.48 and 2.49 is therefore that W can be either positive or negative in eqn 2.48 but only positive in eqn 2.49. Adiabatic work can always be done on the system—say by the paddle in Fig. 2.9, to increase its internal energy—but a system within rigid walls cannot, under adiabatic conditions, return to its original state while conserving energy by performing work on the paddle. The law of conservation of energy is not in itself sufficient to describe thermal processes. A second law of thermodynamics is therefore required, as will be seen in Chapter 3.

When the system is enclosed within adiabatic but non-rigid walls eqn 2.49 is still applicable, but the relationship between pairs of states is rather more complicated, since the coordinates of the system can be varied even when the paddle is not used to perform work on the system. States A and B_1 may then exist which are *mutually* accessible along the reversible adiabatic curve shown schematically in Fig. 2.14. The point B_2 is also accessible from A as the result

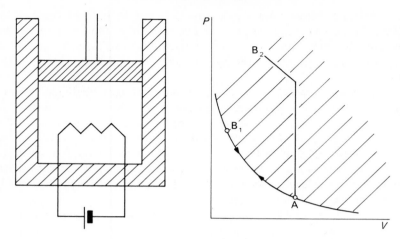

Fig. 2.14 A system in thermal equilibrium within non-rigid adiabatic walls. The system may be able to move from A to B_1 *and* from B_1 to A as the pressure is varied (reversibly) when no electrical work is performed. All the points in the shaded region can be reached from A by a combination of electrical work and change of volume but A cannot be reached from B_2.

of a combination of external work and reversible adiabatic changes. In fact the whole of the shaded region in the figure is accessible from A, but A is not accessible from states within the shaded region.

The general relationship between the states of a system within adiabatic walls is therefore that if A is *not* accessible from B then B is accessible from A but certain pairs of states A and B may be mutually accessible.

In the constant volume process under adiabatic conditions discussed in Section 2.3, $Q = 0$ and

$$W_{\text{ad}} = U(B) - U(A) = \int_{A}^{B} C_V \, dT. \tag{2.50}$$

When the thermal capacity is independent of temperature this becomes simply

$$W_{\text{ad}} = C_V (T_B - T_A). \tag{2.51}$$

Joule's experimental conclusion—that if for a system at constant volume the state B could be reached from A by performing adiabatic work then the state A could *not* be reached from B by performing adiabatic work—is now seen to be equivalent to saying that the temperature of a system within rigid adiabatic walls can be increased by performing work on it but this energy is not recoverable as external work. The system does not *spontaneously* cool back to T_A while performing work on the paddle in Fig. 2.9. This irreversibility of thermal and mechanical processes is the fundamental problem, going beyond the law of conservation of energy, which will be discussed in Chapter 3.

Exercises*

2.1. The resistance of a wire is R_θ at a temperature θ °C measured on the ideal gas scale and

$$R_\theta = R_0(1 + \alpha\theta + \beta\theta^2)$$

where R_0 is the resistance at 0 °C, $\alpha = 3.5 \times 10^{-3}$ (°C)$^{-1}$, $\beta = -3.0 \times 10^{-6}$ (°C)$^{-2}$. Calculate the temperature on the resistance scale when an ideal gas thermometer reads 70 °C.

2.2. The virial equation may be written as in eqns 2.10 or 2.11. Show that
(a) $A = RT$, (b) $B_P = B_V$, (c) $C_V = B_P^2 + AC_P$, (d) $B = b - a/RT$, (e) $C_V = b^2$
where a and b are the van der Waals constants.

2.3. Derive eqns 2.14–2.16.

2.4. Find expressions for the work done by a gas in expanding from volume V_1 to V_2 (a) at constant pressure, and (b) at constant temperature (isothermal expansion) for the following equations of state

(i) $PV = RT$

(ii) $P(V - b) = RT$

(iii) $\left(P + \dfrac{a}{V^2}\right)V = RT$

(iv) $\left(P + \dfrac{a}{V^2}\right)(V - b) = RT.$

Distinguish between external work and work to overcome intermolecular forces.

2.5. Find the change in the internal energy of one mole of a monatomic perfect gas in an expansion from 5 m³ to 10 m³ at a constant pressure of 1 atmosphere (10^5 N m^{-2}) if

$$C_P/C_V = \gamma = 5/3.$$

*Values of the constants required for the exercises are given in Appendixes V and VI.

2.6. Two moles of a monatomic perfect gas are initially at a temperature of 600 K and a pressure of 2 atmospheres. The gas expands reversibly and at constant temperature to twice its original volume. Calculate (a) the work done by the gas, (b) the heat supplied to the gas and (c) the change in the internal energy of the gas.

2.7. A perfect gas is taken through the following cycle: (ab) isothermal expansion at temperature T_1 from volume V_1 to volume $2V_1$, (bc) compression at constant pressure P_1 to volume V_1, (ca) change of pressure at constant volume. Find expressions for the work done, the heat transferred and the change in the internal energy of the gas in each part of the cycle and show that the sum of the three terms is zero for the internal energy only.

2.8. One mole of water at its boiling point is converted into vapour under a constant pressure of 1 atmosphere. The heat absorbed (latent heat) was found to be 41 kJ. Calculate the external work done, assuming that the vapour behaves as a perfect gas. Explain the difference between the heat supplied and the external work done.

2.9. Find an expression for the work done when a wire is heated from temperature T_1 to temperature T_2 at constant tension $\mathscr{F}_0$.

2.10. Show that W is not in general a function of state by considering the consequences of a function $W(T, V)$.

2.11. It should be clear from the discussion in Section 2.1 that a particular thermometer is useful in so far as it either reproduces the perfect gas scale or can be easily corrected to that scale. At low temperature the susceptibility (χ) of a weakly paramagnetic material, Section 10.5, is often used as a thermometer. The equation of state is given by,

$$C\chi^{-1} = T - \theta$$

where T is the temperature on the gas scale and C and θ are constants. Find an expression for the error in the measurement of the temperature of a body at temperature T that would be obtained using this thermometer if it was assumed that θ was zero.

2.12. Show that the infinitesimal work term for a film of liquid of surface tension S and area A is SdA, provided the change in the volume of the film may be neglected.

2.13. Show, by considering the equilibrium of a horizontal layer of gas of thickness dz, density ρ and temperature T, that in a gravitational field,

$$dP = -\rho g \, dz$$
$$P = P_0 e^{-Mgz/RT}$$

where M is the molecular weight of the gas and g the acceleration due to gravity.

2.14. Write down dz for the function

$$z = x^4 y^2 + 5xy^3$$

and show that it satisfies equation A.18.
Establish whether the following expressions are perfect or imperfect differentials and, if necessary, find an integrating factor for the equations,

(a) $3x^2 y + 3y^2) \, dx + (x^3 + 6xy) \, dy$
(b) $x^2 y^2 \, dx + x^3 y \, dy$

2.15. In classical physics the heat capacity of a system at constant volume is independent of temperature. Classical physics however fails at low temperature and the heat capacity of a metal for example may then be written

$$C_v = AT^3 + \gamma T$$

where A and γ are constants, see Sections 10.1 and 10.3. Find an expression for the heat transfer required to raise the temperature of a metal from temperature T_1 to T_2.

2.16. The speed of sound in a gas is given by the expression

$$c_s = \sqrt{\frac{\gamma P}{\rho}}$$

where $\gamma = C_P/C_V$. The value of C_V is $3R/2$ and $5R/2$ for 1 mole of a monatomic and diatomic gas respectively. Show that c_s is only a function of temperature for a perfect gas and find the value of c_s for argon and oxygen at 300 K.

2.17. The Boyle temperature, T_B, is defined as the temperature at which the second virial coefficient, B, is equal to zero. From Exercise 2.2. (d) the value of T_B for a van der Waals gas is a/Rb. At this temperature a real gas will follow the perfect gas law over a wide range of density. The Boyle line is defined as the equation relating the density, ρ, of a gas to the temperature when the perfect gas law holds. Show that for a van der Waals gas the

Boyle line is given by

$$\rho/\rho_c = 3(1 - T/T_B) \tag{1}$$

$$P/P_c = 27(T/T_B)(1 - T/T_B) \tag{2}$$

where ρ_c and P_c are the critical density and pressure. These results are in quite good agreement with experiments. The value of 3 in eqn 1 for example may be compared with experimental values of 3.4 for Ar and 3.7 for N_2. (See J. G. Powles (1983) for further details.)

3

Second law of thermodynamics

The first law of thermodynamics expressed in the form

$$U(B) - U(A) = Q + W \tag{3.1}$$

has been seen to be a form of the law of conservation of energy in which heat is to be treated as a form of energy or, more exactly, as a form of energy *transfer* between systems. The first law is sometimes stated in terms of the impossibility of building a certain kind of perpetual motion machine. The machine must work in a cycle (Section 2.7) and ΔU is zero at the end of each cycle. The first law then requires $W = -Q$ and the efficiency $\eta \leqslant 1$ (eqn 2.40). The efficiency with which heat can be converted into work in a cyclic process cannot, according to the first law of thermodynamics, exceed unity. The fact that real heat engines always have $\eta < 1$ was one of the origins of the second law of thermodynamics.

A thermal process which violates eqn 3.1 cannot occur, but it has already been remarked for the special case of adiabatic processes, that also many changes which do satisfy this equation do not occur. As a further example consider two bodies at slightly different temperatures $T_1 \gtrsim T_2$ inside adiabatic containers which are allowed to exchange energy and are then isolated again. The total internal energy of the two bodies is unchanged and no work is performed so

$$Q_1 + Q_2 = 0. \tag{3.2}$$

The quantity of heat lost by one body is equal to that gained by the other. The first law of thermodynamics does not go beyond this statement. The basis of a calorimeter experiment is that the heat *lost* by the *hot* body (1) is equal to the heat *gained* by the *cold* body (2). Heat flows from hot to cold, or, equivalently, from high to low temperature. The first law fails to *select* the single direction of heat flow between the two bodies.

A further weakness of the first law is that even in the special form for an infinitesimal reversible change it still contains the imperfect differential dQ_R. The first law of thermodynamics is not therefore very useful for performing calculations on the relationships between the measurable parameters of a system. The expression for $(C_P - C_V)$ (eqn 2.34), for example, contains the term $(\partial U/\partial V)_T$ which would be difficult to evaluate by experiment for any system except a gas.

The problem of the imperfect differential dQ_R may be stated mathematically as follows: show that the first law of thermodynamics in the form of equation 2.46 or 2.47 has an *integrating factor* (ϕ^{-1}) such that dQ_R/ϕ is a perfect differential. That is to say the differential of some function of state S defined by

$$\phi \, dS = dQ_R \tag{3.3}$$

where ϕ is some function of the coordinates of the system. Since an integrating factor does not exist in general (Appendix I) for an eqn of the form 2.47 (if $n > 2$) the postulate that ϕ exists must be based on some generalization of physical observation, it is *not* simply a piece of mathematics.

The history of the discovery of the laws of thermodynamics is briefly discussed in Section 3.8. The second law and the function of state S were originally discovered from a consideration of heat engines (Section 3.7). The formulation of the second law in terms of an integrating factor for dQ_R is, however, a more direct approach from which the theorems for heat engines can quickly be deduced as a special case. The establishment of a function of state S is sufficient to simplify the calculation of quantities such as $(C_P - C_V)$ (eqn 3.14), but in order that heat flow and other irreversible processes may be predicted correctly, a rule must also be established for the change of S in such a process. The first law of thermodynamics essentially involves the conservation of energy ($\Delta U = 0$ for a closed system) but the second law will be seen to be the one-sided statement that $\Delta S \geqslant 0$ for a change in a system within adiabatic walls. The direction of heat transfer between bodies therefore involves the selection from the two states which satisfy the first law ($Q_1 + Q_2 = 0$) of the one state for which S is found not to decrease.

The second law of thermodynamics is so important for the development of the subject that an alternative derivation, following the historical approach, is given in an appendix at the end of the chapter, Section 3.10.

3.1 An integrating factor for dQ_R

The first law has been written in eqn 2.47 in the form

$$dQ_R = \sum_{r=1}^{n} Y_r \, dy_r. \tag{3.4}$$

In a reversible adiabatic change ($dQ_R = 0$)

$$\sum_{r=1}^{n} Y_r \, dy_r = 0 \tag{3.5}$$

where Y_r and dy_r are defined by eqn 2.46. Now an integrating factor does not in general exist for an equation of the form of 3.4 if $n > 2$ (Appendix I). The assertion that such an integrating factor exists for the first law is therefore a statement about physics (a physical law) not simply a piece of mathematics. The correct statement of this problem was first given by Carathéodory and it is useful to divide the statement into its mathematical and physical components.

1 *The mathematical theorem of Carathéodory.* An equation of the form 3.4 is integrable if the neighbourhood of any arbitrary point A contains points B inaccessible from A along solution curves of the eqn 3.5.

2 *The physical statement—second law of thermodynamics.* In the neighbourhood of any state A of a system within adiabatic walls there are states B inaccessible from A.

The second statement is a generalisation to points mathematically close to a given point A of the experimental observation of Joule (discussed in Sections 2.3 and 2.8) that for a system within adiabatic walls

$$U(B) - U(A) = W_{ad}$$

but that if state B could be reached from A by performing work on the system, then for a system within rigid adiabatic walls it was *not* possible to return from B to A. When the adiabatic walls were non-rigid, states were still found (Fig. 2.14) from which the state A was inaccessible.

The immediate consequence of these two propositions is that an integrating factor exists for dQ_R. The second law however involves more than this because it is a general statement applicable to both reversible and irreversible changes. The application to irreversible processes will be discussed in Section 3.4 since it is first necessary to establish the function S defined by the equation

$$dQ_R = \phi \, dS \tag{3.6}$$

where ϕ and S are some functions of the parameters of the system. It should immediately be obvious from the form of eqn 3.6 that since ϕ and dS occur as a product there is no unique function ϕ or S. The quantity Q, however, is extensive—proportional to the volume of the system—(see Section 2.8) and therefore ϕ will be chosen to be an intensive quantity so that S is also extensive. The extensive function S will be called the entropy. The intensive integrating factor ϕ then has the remarkable property that it is a function of temperature alone (Appendix I).

$$\phi = F(\theta) \tag{3.7}$$

where $F(\theta)$ is called the absolute temperature function since $F(\theta)$ is the integrating factor for *any* system at temperature θ.

The integrating factor for any system at temperature T, where T is defined on the perfect gas scale, will now be shown to be simply T, that is to say the thermodynamic and perfect gas scales are identical.

The first law of thermodynamics for a perfect gas is written as in eqn 2.31

$$dQ_R = \left(\frac{\partial U}{\partial T}\right)_V dT + P\, dV$$

The equation of state of the perfect gas is $PV = RT$ for one mole and the internal energy is a function only of the temperature.

Consider ϕ as some function of temperature on the perfect gas scale (T) such that

$$\phi^{-1} = f(T).$$

Then the second law of thermodynamics may be written

$$dS = f(T)\left(\frac{\partial U}{\partial T}\right)_V dT + f(T)P\, dV.$$

S is a function of state of, in this case, only two variables so

$$\frac{\partial}{\partial V}\left[f(T)\left(\frac{\partial U}{\partial T}\right)_V\right]_T = \frac{\partial}{\partial T}[f(T)P]_V \tag{3.8}$$

(Appendix I). The left-hand side of eqn 3.8 is zero since both $f(T)$ and $(\partial U/\partial T)_V$ are independent of V. The right-hand side is

$$P\left(\frac{df(T)}{dT}\right) + f(T)\left(\frac{\partial P}{\partial T}\right)_V = P\frac{df(T)}{dT} + \frac{P}{T}f(T) = 0.$$

Therefore

$$\frac{df(T)}{f(T)} = -\frac{dT}{T}$$

and

$$\ln f(T) = -\ln T + \text{constant}$$

where

$$T = \frac{c}{f(T)} = c\phi. \tag{3.9}$$

The thermodynamic and perfect gas scales are therefore linearly related and if both scales are defined to be 273.16 K at the triple point of water, they will agree at all other temperatures (Section 2.1). An expression equivalent to

$$dQ_R = T\, dS$$

was first derived by Clausius using the Carnot heat-engine cycle (Section 3.7, and Section 3.10).

3.2 Entropy as a function of state

The second law of thermodynamics can now be written using eqns 3.4, 3.6, and 3.9

$$T\,dS = \sum_{r=1}^{n} Y_r\,dy_r \qquad (3.10)$$

in general and for a simple system with only two independent variables (from eqn 2.33)

$$T\,dS = C_V\,dT + \left[P + \left(\frac{\partial U}{\partial V}\right)_T \right]dV. \qquad (3.11)$$

The eqns 3.10 and 3.11 are correct for either a reversible or an irreversible change between equilibrium states A and B, since if S is a function of state, $S(B) - S(A)$ is independent of the process by which the system arrived at B from A. Only for a reversible change however does $dS = dQ_R/T$. Reversible changes are considered in Section 3.3 and irreversible changes in Section 3.4.

The fact that S is a function of state may be used to simplify eqn 3.11. The equation may be written

$$dS = \frac{1}{T}\left(\frac{\partial U}{\partial T}\right)_V dT + \frac{1}{T}\left[P + \left(\frac{\partial U}{\partial V}\right)_T \right]dV.$$

Then

$$\left(\frac{\partial}{\partial V}\right)\left[\frac{1}{T}\left(\frac{\partial U}{\partial T}\right)_V\right]_T = \frac{\partial}{\partial T}\left\{\frac{1}{T}\left[P + \left(\frac{\partial U}{\partial V}\right)_T \right]\right\}_V \qquad (3.12)$$

therefore

$$\frac{1}{T}\frac{\partial^2 U}{\partial V\,\partial T} = \frac{1}{T}\left[\left(\frac{\partial P}{\partial T}\right)_V + \frac{\partial^2 U}{\partial T\,\partial V}\right] - \frac{1}{T^2}\left[P + \left(\frac{\partial U}{\partial V}\right)_T \right]$$

or

$$P + \left(\frac{\partial U}{\partial V}\right)_T = T\left(\frac{\partial P}{\partial T}\right)_V \qquad (3.13)$$

since

$$\frac{\partial^2 U}{\partial V\,\partial T} = \frac{\partial^2 U}{\partial T\,\partial V}.$$

The important result of this calculation which depends only upon S being some (unknown) function of state is that an eqn (3.13) has been established from which $(\partial U/\partial V)_T$ may be *calculated* for any simple system if the equation of state is known.

The equation

$$\left(\frac{\partial U}{\partial V}\right)_T = T\left(\frac{\partial P}{\partial T}\right)_V - P = T^2\frac{\partial}{\partial T}\left(\frac{P}{T}\right)_V \qquad (3.14)$$

is called the *energy equation*. The second form of eqn 3.14 shows that $(\partial U/\partial V)_T$ is zero if the equation of state has the form, P/T is constant at constant volume, or in general

$$P = f(V) T.$$

This is of course the form of the perfect gas equation $PV = RT$ which, substituted into eqn 3.14, does indeed lead to $(\partial U/\partial V)_T = 0$ for a perfect gas.

The second law of thermodynamics for a simple system (eqn 3.11) can therefore be written

$$T \, dS = C_V \, dT + T \left(\frac{\partial P}{\partial T} \right)_V dV \qquad (3.15)$$

or equivalently, since S may be considered as a function of T and V

$$T \, dS = T \left(\frac{\partial S}{\partial T} \right)_V dT + T \left(\frac{\partial S}{\partial V} \right)_T dV. \qquad (3.16)$$

A comparison of eqn 3.15 with 3.16 establishes the important results

$$C_V = T \left(\frac{\partial S}{\partial T} \right)_V \qquad (3.17)$$

$$\left(\frac{\partial P}{\partial T} \right)_V = \left(\frac{\partial S}{\partial V} \right)_T. \qquad (3.18)$$

The entropy change of a system at constant volume is from eqn 3.15

$$S(T_B, V) - S(T_A, V) = \int_{T_A}^{T_B} \frac{C_V}{T} dT = \int_{T_A}^{T_B} C_V \, d(\ln T) \qquad (3.19)$$

the area under a graph of C_V against $\ln T$ (remember that in general C_V is a function of both temperature and volume).

The condition for C_V to be independent of V may be established from eqn 3.17

$$\left(\frac{\partial C_V}{\partial V} \right)_T = \frac{\partial}{\partial V} \left[T \left(\frac{\partial S}{\partial T} \right)_V \right]_T$$

$$= T \frac{\partial}{\partial V} \left(\frac{\partial S}{\partial T} \right)_V$$

$$= T \frac{\partial}{\partial T} \left(\frac{\partial S}{\partial V} \right)_T$$

on changing the order of differentiation. Then using eqn 3.18

$$\left(\frac{\partial C_V}{\partial V} \right)_T = T \frac{\partial}{\partial T} \left(\frac{\partial P}{\partial T} \right)_V = T \left(\frac{\partial^2 P}{\partial T^2} \right)_V. \qquad (3.20)$$

The condition for the heat capacity of any system to be independent of volume is therefore that $(\partial^2 P/\partial T^2)_V$ be zero, a condition satisfied by both a perfect gas and a van der Waals gas.

The four important results given in eqns 3.14, 3.17, 3.18, and 3.20 have been established simply by using the fact that S is a function of state and the general mathematical relations between partial differentials. The nature of S (the 'meaning' of entropy) is not involved. The results could not be obtained from the first law of thermodynamics which contains the imperfect differential dQ_R

Since any function of functions of state is also itself a function of state there are many relations of the form 3.18 (known as Maxwell relations) some of which are derived in Section 4.2. Their importance of course is that quantities such as $(\partial S/\partial V)_T$ can always be removed from equations in which they occur and replaced by differentials which relate to quantities that are more accessible to experiment.

The function $(\partial P/\partial T)_V$ for example may be measured easily for a gas but would be difficult to measure for a solid or liquid. However using the relation

$$\left(\frac{\partial P}{\partial T}\right)_V \left(\frac{\partial T}{\partial V}\right)_P \left(\frac{\partial V}{\partial P}\right)_T = -1 \tag{3.21}$$

(Appendix I), and the definitions of the isothermal compressibility (κ_T) and coefficient of volume expansion (α)

$$\kappa_T = -\frac{1}{V}\left(\frac{\partial V}{\partial P}\right)_T = -\left(\frac{\partial \ln V}{\partial P}\right)_T \tag{3.22}$$

$$\alpha = \frac{1}{V}\left(\frac{\partial V}{\partial T}\right)_P = \left(\frac{\partial \ln V}{\partial T}\right)_P . \tag{3.23}$$

Therefore

$$\left(\frac{\partial S}{\partial V}\right)_T = \left(\frac{\partial P}{\partial T}\right)_V = \alpha\kappa_T^{-1} = \alpha B_T \tag{3.24}$$

It should be remembered that α and κ_T are in general functions of temperature and volume. (The reciprocal of κ_T is called the isothermal bulk modulus.)

These two simple calculations illustrate one of the most important features of thermodynamics. The number of *independent* thermodynamic quantities for a system is limited by the form of the second law due to the relationships of the type shown in eqns 3.12 or 3.24 which must exist between the different terms if S is to be a function of state. The manipulation of the mathematical properties of the differentials of a function of n variables ($n=2$ for a simple system) therefore leads to relationships for quantities such as $(C_P - C_V)$ or $(\partial S/\partial V)_T$ in terms of other quantities which are possibly easier to measure experimentally.

Thermodynamics cannot be used to establish the magnitude of any single physical quantity although it can, in certain special cases, show that they become zero or infinite. C_V and α for example tend to zero as the temperature approaches absolute zero (Section 4.4) and since $(\partial P/\partial V)_T$ is zero at the critical point (Section 2.2) the isothermal compressibility goes to infinity as the temperature approaches T_c (see also Section 10.4).

3.3 The calculation of entropy changes in principle

The calculation of the change of entropy of a system between two states A and B is straightforward in principle.

$$\Delta S = S(B) - S(A) = \int_A^B \frac{\mathrm{d}Q_R}{T}$$

where the integral is independent of the reversible path since S is a function of state.

In particular, since the entropy is a function of state

$$\oint \frac{\mathrm{d}Q_R}{T} = 0.$$

The entropy was first introduced in terms of $\mathrm{d}S$ (eqn 3.6) so there is no meaning for absolute entropy in thermodynamics but the change in entropy $S(B) - S(A)$ is well defined and independent of the path joining A and B.

The simplest entropy calculation is for a process occurring at constant volume and temperature. A thermal reservoir (Section 2.7) at temperature T_R for example can gain or lose a quantity of heat Q without change of temperature. The entropy change of the reservoir is then simply

$$\Delta S = \frac{Q}{T_R}. \tag{3.25}$$

Similarly unit mass of a solid at its melting temperature (T_m) requires a further quantity of heat (the latent heat l) to convert it to liquid and

$$\Delta S = \frac{l}{T_m}. \tag{3.26}$$

In general the entropy must be considered as a function of two variables (for a simple system) say volume and temperature (as in eqn 3.19)

$$S(V_B, T_B) - S(V_A, T_A) = \int_{T_A}^{T_B} \frac{C_V \, \mathrm{d}T}{T} + \int_{V_A}^{V_B} \left(\frac{\partial P}{\partial T}\right)_V \mathrm{d}V. \tag{3.27}$$

When C_V is a function of V it is necessary to integrate the first term on the right-hand side of the equation at constant volume (say V_A) and then to evaluate $(\partial P/\partial T)_V$ at temperature T_B. The result for ΔS would of course be just the same if C_V were to be considered at volume V_B and $(\partial P/\partial T)_V$ evaluated at T_A since the entropy change is independent of the path (Fig. 3.1).

In the particularly simple case of a system where C_V is independent of the volume, such as a perfect or van der Waals gas, the integration is more

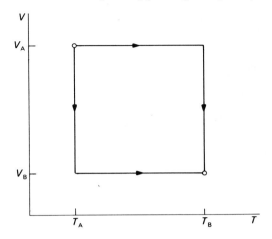

Fig. 3.1 Two possible paths for the integration of eqn 3.27 when C_V is a function of both temperature and volume.

straightforward. The specific heat of a monatomic perfect gas is independent of volume *and* temperature so C_V may be removed from the integral and

$$S(V_B, T_B) - S(V_A, T_A) = C_V \ln \frac{T_B}{T_A} + R \ln \frac{V_B}{V_A} \qquad (3.28)$$

on substitution of R/V for $(\partial P/\partial T)_V$ in the second integral of eqn 3.27. The behaviour of this equation as T_A goes to zero indicates that a classical perfect gas cannot exist at absolute zero if the entropy is to remain finite. This is discussed in Section 4.4. Further examples of entropy changes for particular processes will be considered in Section 3.5 and in the exercises at the end of the chapter but in this section we shall concentrate on the principles involved in such calculations.

The entropy change of a system between the states of thermodynamic equilibrium A and B, $S(B) - S(A)$, is independent of the process by which the system goes from A to B, whether by any reversible path or by an irreversible process, since S is a function of state. The *calculation* of the entropy change in either case is made by finding any convenient reversible path which joins A to B.

As an example of an irreversible process, consider an isolated volume of a perfect gas (Fig. 3.2) separated from a vacuum by an adiathermal wall. When the wall is removed the gas will flow turbulently (irreversibly) into the vacuum and after some time come to a new equilibrium state. Since U is a function only of T for a perfect gas, no net work was performed and no heat transferred, the temperature of the gas is unchanged. The initial state is $(T,$

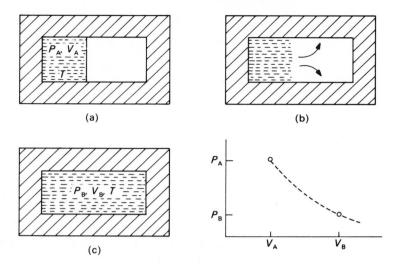

Fig. 3.2 (a) A perfect gas in thermal equilibrium with volume V_A and temperature T. (b) The gas flows via non-equilibrium states into the vacuum. (c) The final equilibrium state. The irreversible process is replaced by any reversible process joining A to B in order to calculate the change of entropy.

V_A) and the final state (T, V_B). The intermediate states are not defined since the system was not in thermal equilibrium. The entropy change of the gas may be calculated by finding a reversible process which connects the two states. The simplest such process is a reversible isothermal change from V_A to V_B (Fig. 3.2). Then from eqn 3.28

$$S(V_B,\ T) - S(V_A,\ T) = R \ln\left(\frac{V_B}{V_A}\right). \tag{3.29}$$

The entropy of the system has *increased* after the irreversible expansion. Now Q was zero in this process, since the entire system was surrounded by adiabatic walls, but the entropy has increased. The equation

$$\mathrm{d}S = \frac{\mathrm{d}Q_R}{T}$$

for reversible processes must therefore be extended to read

$$\mathrm{d}S \geqslant \frac{\delta Q}{T} \tag{3.30}$$

where the equality sign is restricted to reversible processes. In an irreversible process the temperature T in eqn 3.30 is the temperature of the thermal reservoir which supplies heat to the system, since the temperature of a nonequilibrium system is not defined. The increase in the entropy of a system due to an irreversible process will now be considered. (The student could usefully work Exercise 3.2 before reading further.)

3.4 Principle of increase of entropy

The statement of the second law of thermodynamics given in Section 3.1 'in the neighbourhood of any state A of a system within adiabatic walls there are states B inaccessible from A' was stated to be true for both reversible and irreversible processes. A consideration of the first law of thermodynamics for reversible processes then led to the identification of a function of state, the entropy S, such that

$$dS = \frac{dQ_R}{T}.$$

In any reversible adiabatic process dQ_R is zero and hence the entropy of the system is constant. The system therefore may be considered to move on the *surface* defined by

$$S(Y, y) = \text{constant}$$

in a reversible adiabatic process. In a simple system with only two independent variables the surface reduces to a line in the $T-V$ space given for example by the equation $TV^{(\gamma-1)} = \text{constant}$ for a perfect gas (eqn 3.45). A reversible adiabatic process is sometimes called an *isentropic* process since it takes place at constant entropy.

The essential point in the argument for an irreversible process is to note that A must be one of the end points of the region of all those states which are *accessible* from A or else there will be *no* neighbouring state B which is inaccessible as required by the second law. This process is shown schematically in Fig. 3.3. The entropy of a system after an adiabatic change can therefore only stay constant or change in one direction. Since in Section 3.3 the irreversible expansion of a gas was seen to lead to an *increase* in the entropy it is clear that the entropy must always stay constant or increase in an adiabatic process. The condition that the entropy increase (rather than decrease) in fact follows from the definition of the thermodynamic temperature as a positive quantity.

The principle of the increase of entropy may be stated in the form 'the final state (B) of a system within adiabatic walls is never less than the initial state (A)'

$$S(B) - S(A) \geqslant 0 \quad \text{(adiabatic)} \tag{3.31}$$

The problem of the direction of heat flow discussed at the beginning of Chapter 3 is now solved. The bodies 1 and 2 exchange heat such that

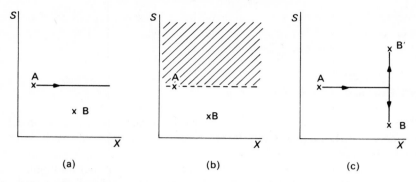

Fig. 3.3 A schematic illustration of the second law of thermodynamics. (a) In a reversible adiabatic process the entropy remains constant (an isentropic process). The point B is inaccessible from A. (b) In an irreversible adiabatic process the entropy can only change in one direction (increase). The point B is therefore again inaccessible. (c) If the region accessible from A included both points of increased and decreased entropy a combination of irreversible and reversible processes would make all points B accessible. The second law of thermodynamics would therefore not be satisfied.

$Q_1 + Q_2 = 0$ (eqn 3.2) and we take T_1 to be slightly greater than T_2. Then

$$\Delta S = \frac{Q_1}{T_1} + \frac{Q_2}{T_2} = Q_2 \left(\frac{1}{T_2} - \frac{1}{T_1} \right). \tag{3.32}$$

The requirement that $\Delta S \geqslant 0$ then leads to $Q_2 > 0$. The cooler body gains heat from the warmer body. Since ΔS increases in eqn 3.32, *heat flow across a finite temperature difference is an irreversible process.*

If the requirement that T_1 be close to T_2 is relaxed then the entropy change must be calculated by finding a *reversible* path joining initial and final states. The reversible path is formed by placing each body separately in contact with a series of thermal reservoirs whose temperatures vary continuously between the initial and final temperatures. In the most general case of bodies 1 and 2 with heat capacities C_{V1} and C_{V2} (both functions of temperature) and initial temperatures T_{1A} and T_{2A} $(T_{1A} > T_{2A})$ the final temperature T_B is given by the calorimeter equation

$$\int_{T_{1A}}^{T_B} C_{V1} \, dT = - \int_{T_{2A}}^{T_B} C_{V2} \, dT \tag{3.33}$$

and the total entropy change by

$$\Delta S = \int_{T_{1A}}^{T_B} \frac{C_{V1} \, dT}{T} + \int_{T_{2A}}^{T_B} \frac{C_{V2} \, dT}{T} \tag{3.34}$$

$$= \int_{T_{1A}}^{T_B} C_{V1} \, d(\ln T) + \int_{T_{2A}}^{T_B} C_{V2} \, d(\ln T). \tag{3.35}$$

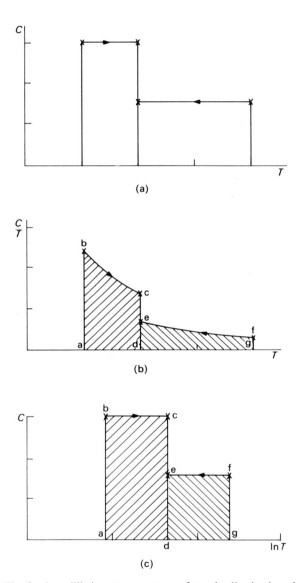

Fig. 3.4 (a) The final equilibrium temperature of two bodies is given by equal areas under the graph of heat capacity against temperature (eqn 3.33). (In the figure the heat capacities have been taken to be independent of temperature for simplicity.) (b) The total entropy change of the two bodies is given by the difference between the areas abcd and defg when C/T is plotted against the absolute temperature (T) or (c) when C is plotted against ln T.

These equations are represented in Fig. 3.4. ΔS is always positive. In the particularly simple case $C_{V1} = C_{V2} = C_V$ independent of T, $T_B = (T_{1A} + T_{2A})/2$, the integration is straightforward

$$\Delta S = C_V \left(\ln \frac{T_B}{T_{1A}} + \ln \frac{T_B}{T_{2A}} \right)$$

which is positive definite (remember that the first term is negative since $T_{1A} > T_B > T_{2A}$). In the limit $T_{1A} = (T_{2A} + \Delta)$, the reader may prove using $\ln(1 \pm y) \approx \pm y - y^2/2$ for small y that

$$\Delta S = + \frac{C_V \Delta^2}{4 T_B^2} \qquad (\Delta \to 0) \tag{3.36}$$

3.5 The entropy of a perfect gas

The change of entropy of a perfect gas from the initial state (V_A, T_A) to the final state (V_B, T_B) was shown to be

$$\Delta S = S(V_B, T_B) - S(V_A, T_A)$$

$$\Delta S = \int_{T_A}^{T_B} \frac{C_V \, dT}{T} + \int_{V_A}^{V_B} \left(\frac{\partial P}{\partial T} \right)_V dV \,.$$

(eqn 3.27). The thermal capacity of a monatomic gas is independent of temperature and therefore (eqn 3.28)

$$\Delta S = C_V \ln \left(\frac{T_B}{T_A} \right) + R \ln \left(\frac{V_B}{V_A} \right). \tag{3.37}$$

The perfect gas is so important in thermal physics— because of its particularly simple equation of state $(PV = RT)$ and the fact that the internal energy is independent of the volume—that it is worth considering eqn 3.37 in more detail. It has already been emphasised that since the entropy was introduced by a differential $(dS = dQ_R/T)$ there is no absolute value of the entropy, only entropy changes are completely defined. If the state (V_A, T_A) were to be taken as a standard reference state however, eqn 3.37 might be written

$$S(V, T) = C_V \ln T + R \ln V + S_0 \tag{3.38}$$

$$S_0 = S(V_A, T_A) - C_V \ln T_A - R \ln V_A.$$

Equation 3.38 is of course correct and is often found in textbooks, but it is rather misleading. The appearance of $\ln T$ and $\ln V$ is mathematical nonsense unless the definition of S_0 is remembered, but more seriously the entropy as defined by eqn 3.38 is *non-extensive*.

This becomes obvious if the heat capacity C_V is replaced by the specific heat per molecule c_V and the number of molecules N and the gas constant replaced by $N k$ where k is the Boltzmann constant.

$$S(V, T) = N c_V \ln T + N k \ln V + S_0. \tag{3.39}$$

If the volume of the system is divided by two, the first term on the right is also halved, since T is constant throughout the volume, but the second term becomes $(N k/2) \ln(V/2)$. The logarithmic terms must therefore both be made intensive if the entropy is to be proportional to N.

This problem can be overcome if the reference state is defined in terms of the intensive parameters (P_A, T_A). Then using $P_A V_A = N k T_A$ eqn 3.39 becomes

$$S(V, T) = N c_V \ln T + N k \ln \left(\frac{V}{N} \right) + N s_0 \tag{3.40}$$

$$N s_0 = - N c_V \ln T_A - N k \ln \left(\frac{k T_A}{P_A} \right)$$

$$= - N c_V \ln T_A - N k \ln \left(\frac{V_A}{N} \right). \tag{3.41}$$

Now N/V, the number of particles per unit volume, is constant throughout the system since gravitational forces have not been considered, so $n = N/V$ is an intensive quantity. The entropy is therefore properly extensive (proportional to N) in eqn 3.40. The difference between eqns 3.39 and 3.40 may seem to be rather trivial, but a great deal of confusion can arise if calculations are performed in a manner which does not make the extensive nature of the entropy obvious.

Consider for example two equal volumes of a perfect gas at the same temperature and pressure in adiabatic enclosures (Fig. 3.5a). What is the change in the entropy of the system if the partition between the two volumes of gas is withdrawn reversibly? The answer is clearly zero since nothing has changed in the gas after the removal of the partition and this is confirmed immediately by eqn 3.40. The temperature of the two volumes of gas (T) is unchanged and if the number of molecules in each container was $N/2$ and the volume $V/2$ then the original entropies of the two volumes may be written

$$S_{A1} = S_{A2} = \frac{N}{2} c_V \ln T + \frac{N}{2} k \ln \left(\frac{V}{N} \right) + \frac{N}{2} s_0$$

the final entropy

$$S_B = N c_V \ln T + N k \ln \left(\frac{V}{N} \right) + N s_0$$

and the entropy change

$$\Delta S = S_B - (S_{A1} + S_{A2}) = 0. \tag{3.42}$$

The partition could of course equally well be inserted reversibly into the large volume of Fig. 3.5b again without any change in the entropy. The important

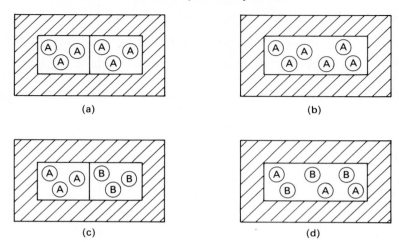

Fig. 3.5 (a) The entropy of two equal volumes of a single gas at the same temperature and pressure is unchanged after the partition is removed from within the adiabatic enclosure (b), because the number density of the molecules is unchanged. (c) The entropy increases if two different gases are mixed (d).

parameter is the number density of the molecules (N/V) not the number or the volume separately, as is clearly shown by eqn 3.40. A further discussion of this experiment, (from the microscopic point of view) is given in Section 7.5. The important point that then becomes clear is that the entropy change is zero because the molecules of the gas are indistinguishable, but this point has already really been established by the above equations which are concerned only with (N/V) and not with which particular molecules of the gas make up N in any given V.

When two different perfect gases are contained in equal volumes at the same temperature and pressure within adiabatic walls (Fig. 3.5c) the entropy does increase after the partition has been removed since (N/V) for each type of molecule decreases by a factor of two. The irreversible mixing of the two gases has to be replaced by a reversible process in order to calculate the entropy change. It was first shown by Gibbs that the entropy change could be calculated by considering the movement across the total volume of a membrane which allowed molecules say of gas 1 but not of gas 2 to pass through it. The mixture shown in Fig. 3.5d could therefore be separated as in Fig. 3.5c. The overall effect however is simply equivalent to the free expansion of each gas from a volume $V/2$ to a volume. V. The molecules of a perfect gas essentially ignore each other. Then writing the number of molecules of type 1 as N_1

$$S_{A1} = N_1 c_V \ln T + N_1 k \ln\left(\frac{V}{2N_1}\right) + N_1 s_0$$

$$S_{A2} = N_2 c_V \ln T + N_2 k \ln \left(\frac{V}{2N_2}\right) + N_2 s_0'$$

$$S_B = (N_1 + N_2)c_V \ln T + N_1 k \ln \left(\frac{V}{N_1}\right)$$

$$+ N_2 k \ln \left(\frac{V}{N_2}\right) + N_1 s_0 + N_2 s_0'.$$

Since the number of molecules of type 1 is equal to the number of type 2 the entropy change may be written using $N_1 = N_2 = N/2$

$$\Delta S = S_B - (S_{A1} + S_{A2}) = (N_1 + N_2)k \ln 2 = N k \ln 2. \tag{3.43}$$

The entropy of mixing of two gases is therefore positive (an irreversible process) as expected since reinserting the partition in Fig. 3.5d does not lead to the original state of the system.

3.6 Adiabatic equation for a perfect gas

The equation of the path of a perfect gas in a reversible adiabatic process (a process at constant entropy or an isentropic process) may be calculated from the general form of the second law of thermodynamics for a perfect gas.

$$T \, dS = C_V \, dT + P \, dV.$$

In an isentropic process dS is zero and

$$C_V \, dT = -P \, dV = -\frac{RT}{V} dV$$

$$\int \frac{C_V \, dT}{T} = -R \int \frac{dV}{V}.$$

When C_V is independent of temperature (a monatomic gas) the equation becomes simply

$$\int \frac{dT}{T} = -\frac{R}{C_V} \int \frac{dV}{V}$$

or

$$TV^{R/C_V} = \text{constant.} \tag{3.44}$$

Since $R = (C_P - C_V)$ (eqn 2.35), eqn 3.44 becomes

$$TV^{(\gamma-1)} = \text{constant} \tag{3.45}$$

where $\gamma = C_P/C_V$. Since all the states of thermal equilibrium of the gas are related by the equation $PV/T = \text{constant}$, eqn 3.45 may also be written

$$PV^\gamma = \text{constant} \tag{3.46}$$

or

$$P^{(1-\gamma)} T^\gamma = \text{constant}. \tag{3.47}$$

These equations are required in the study of the Carnot cycle for a perfect gas (Section 3.7) and are also important since a measurement of γ provides some evidence for the number of atoms in a molecule of a gas (Section 7.5).

3.7 The Carnot theorems for heat engines

A heat engine was introduced in Section 2.7 as a machine which after each complete cycle has converted some quantity of heat into useful work. The maximum efficiency of such a machine can now be established using the statement that the entropy of a system within adiabatic walls cannot decrease (Section 3.4). Consider first a heat engine working between two large thermal reservoirs at temperatures T_A and T_B ($T_A > T_B$). Then if the engine is to be reversible all heat must be taken in at T_A and rejected at T_B to avoid heat transfer across a temperature gradient. A cycle therefore consists of two isothermals at T_A and T_B joined by two reversible adiabatic changes (Fig. 3.6). The efficiency of the engine is given by eqn 2.39 as

$$\eta = \frac{Q_A - Q_B}{Q_A} \tag{3.48}$$

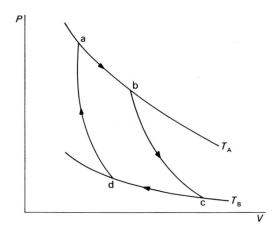

Fig. 3.6 The only reversible cycle for an engine working between two infinite heat reservoirs at T_A and T_B consists of a reversible isothermal expansion at T_A, a reversible adiabatic expansion to temperature T_B, an isothermal contraction at T_B and a final adiabatic contraction. In this way no heat is transferred across a finite temperature difference. The shape of the P–V diagram will depend upon the working substance, here a perfect gas, but the efficiency of the engine is independent of the working substance.

where $Q_A(Q_B)$ is the magnitude of the quantity of heat absorbed (rejected) by the engine at temperature $T_A(T_B)$. After a complete cycle, ΔU is zero, so the overall entropy change for the system of engine plus reservoirs is

$$\Delta S = -\frac{Q_A}{T_A} + \frac{Q_B}{T_B}.$$

In a reversible process ΔS is zero and

$$\frac{Q_A}{Q_B} = \frac{T_A}{T_B} \quad \text{(reversible).} \tag{3.49}$$

Hence the efficiency of a reversible engine working between maximum temperature T_A and minimum temperature T_B is

$$\eta = \frac{Q_A - Q_B}{Q_A} = \frac{T_A - T_B}{T_A} \quad \text{(reversible).} \tag{3.50}$$

In general $\Delta S \geqslant 0$ and therefore

$$\frac{Q_A}{Q_B} \leqslant \frac{T_A}{T_B}$$

$$\eta \leqslant 1 - \frac{T_B}{T_A} \tag{3.51}$$

where the equality is correct for a reversible engine. A *reversible* engine is therefore the most efficient possible engine working between a maximum temperature T_A and a minimum temperature T_B but always has an efficiency of less than unity since $T_A > 0$ K(Section 4.4).

The concept of a reversible heat engine was introduced by Carnot (1824) and from the Carnot theorems Kelvin and Clausius first formulated the second law of thermodynamics (Sections 3.8 and 3.10). The historical order has been inverted in this book because the second law of thermodynamics applies to all systems and the Carnot theorems may then be treated as special applications of the law to heat engines.

The Carnot theorems may be written:

(1) all reversible engines working between two given heat sources have the same efficiency, independent of the nature of the working substance;
(2) a reversible engine is always more efficient than an irreversible engine working between the same source and sink.

The first theorem follows immediately since no assumption about the working substance (perfect gas, real gas, vapour, etc.) was made in deriving eqn 3.50 and the second theorem follows from eqn 3.51.

A reversible heat engine is often called a Carnot engine and a reversible cycle of two isotherms and two adiabats a Carnot cycle. This terminology has however led to some confusion and the term reversible heat engine will be used in this book. When the engine works between only *two* reservoirs, the only possible reversible engine is one which follows a Carnot cycle because the engine must absorb heat at T_A and reject it at T_B to prevent a heat transfer across a finite temperature gradient. A graph of temperature against entropy for a Carnot cycle is simply a rectangle independent of the nature of the working substance since two changes occur at constant temperature and two at constant entropy (Fig. 3.7). A Carnot cycle can also be illustrated on a P–V diagram (Fig. 3.6) but the shape of the cycle is then a function of the working substance. A direct verification of eqn 3.50 may be obtained by considering a perfect gas as the working substance (Exercise 3.4).

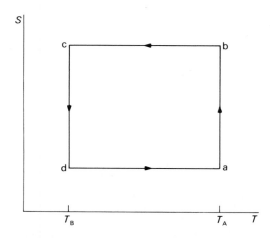

Fig. 3.7 The entropy–temperature diagram of a reversible heat engine working between two thermal reservoirs at T_A and T_B is always a rectangle since the temperature is constant for two parts of the cycle and the entropy for the other two. The area of the rectangle gives the net heat absorbed in the cycle (Exercise 3.6).

When a set of temperature reservoirs are available at temperatures between T_A and T_B many reversible cycles become possible. The Carnot cycle is in fact not of any practical importance because the work done per cycle (the area enclosed by the cycle on a P–V diagram) is very small. An important practical cycle is due to Stirling. The Carnot and Stirling cycles, using a perfect gas as the working substance, are shown in Fig. 3.8. The area of the Stirling cycle is obviously much greater than that of the Carnot cycle under the same limiting conditions (Exercise 3.4) but the efficiencies are identical since the Stirling engine absorbs more heat along the isotherm T_A than does the Carnot engine. The Stirling engine rejects heat to a thermal source of *variable* temperature along bc such that the working substance and source are at the same

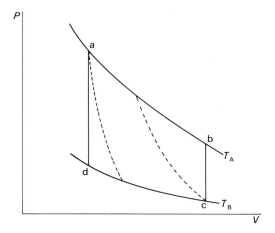

Fig. 3.8 When more than two thermal reservoirs are available many reversible cycles become possible. The Stirling cycle is of great practical importance because the work done per cycle is greater than that of the Carnot cycle (shown as a broken line) under the same conditions. The heat absorbed along da in the Stirling cycle is rejected along bc. The heat is always transferred to a reservoir at the same temperature as the gas. (See also Exercise 3.4).

temperature and absorbs an equal amount of heat from the source along da. The cycle is therefore reversible since heat is never transferred across a finite temperature gradient.

The Stirling cycle has the same efficiency as the Carnot cycle because at the end of the cycle the net effect is that heat has been taken in at the highest temperature and rejected at the lowest temperature. Some reversible cycles, such as the Otto cycle (Exercise 3.13), do not satisfy this criterion and therefore have an efficiency lower than that of a Carnot cycle working between the same maximum and minimum temperatures.

The simple equation for the efficiency of a heat engine (3.51) is one of the most remarkable in all science. An engine designer will normally have the lowest temperature fixed by the environment of the engine, say river water or the atmosphere, and then has only two variables available, the highest temperature which the materials of the engine will withstand (the metallurgical limit) and the degree of irreversibility of the engine due to friction, turbulence in the working substance and heat transfer across a finite temperature difference. All real engines are irreversible with efficiencies at best about 40 per cent of a reversible engine working under the same conditions. The approximate P–V diagram for a real Stirling engine is illustrated in Fig. 3.9.

A *refrigerator* is a machine which, acting in a cycle, extracts heat from a cold body and rejects heat at a higher temperature (Fig. 3.10). It is of course necessary to do work *on* the refrigerator with a separate engine. The work

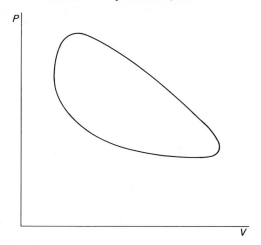

Fig. 3.9 An approximate $P - V$ diagram for a real Stirling engine. The efficiency is found to be about 40 per cent of the theoretical maximum efficiency.

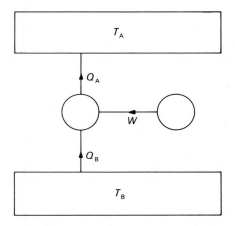

Fig. 3.10 A refrigerator takes heat from a cold body (B) to a hot body (A) with the aid of an external engine.

required is

$$W = Q_A - Q_B$$

where Q_A is the heat rejected to the hot reservoir and Q_B the heat absorbed from the cold reservoir. The coefficient of performance of a refrigerator (CP) is defined

$$CP = \frac{\text{heat extracted from cold reservoir}}{\text{work done on the system}}$$

$$= \frac{Q_B}{Q_A - Q_B}. \tag{3.52}$$

In the case of a reversible refrigerator, eqn 3.49 may be used to give

$$CP = \frac{T_B}{T_A - T_B} \tag{3.53}$$

Notice that the coefficient of performance in eqn 3.53 is *greater* than unity. A reversible engine has of course the highest possible coefficient of performance for a given (T_A, T_B) as may be proved by the same argument by which the reversible heat engine was shown to be most efficient. A refrigerator working in the Stirling cycle (driven by an electric motor) has become of great importance in the liquefaction of gases (Section 5.3).

A *heat pump* is defined as a machine, working in a cycle, which provides useful heating (say of a room) by extracting heat from a thermal reservoir (outside air) at a lower temperature. The process therefore looks like that of the refrigerator shown in Fig. 3.10 but it is the heat given to the high-temperature reservoir which is now of interest. The *coefficient of performance* for a heat pump is

$$\frac{\text{heat rejected to hot reservoir}}{\text{work done on the system}} = \frac{Q_A}{Q_A - Q_B}$$

$$= 1 + CP \tag{3.54}$$

(using eqn 3.52 for a refrigerator). A reversible engine is therefore as usual most efficient since CP is a maximum for a reversible engine.

Lord Kelvin first noticed that a reversible heat pump was the most efficient possible method of converting energy (say electricity) into heat but although some large buildings have been heated by heat pumps the capital cost of the installation has usually more than balanced the saving in running expenses over more conventional heating systems.

3.8 History of thermodynamics

The view of thermodynamics which has been presented in this chapter and the previous one has taken little account of the historical development of the subject. (A more traditional treatment is given in Section 3.10.) A scientific theory very rarely develops as straightforwardly as it is presented in a textbook, but the history of thermodynamics is perhaps the most complicated of any branch of physics. Although no attempt will be made to cover the historical background in detail, it is of some interest to see with the advantage of hindsight exactly where the confusion arose that so hindered the development of thermodynamics.

In Chapter 2 heat was treated as a *derived* quantity such that the change in the internal energy between two states was always given by

$$U(\text{B}) - U(\text{A}) = Q + W.$$

The internal energy was recognised as a function of state and W, and therefore Q, were shown not to be functions of state.

Although the nature of the energy transfer due to thermal processes does not therefore concern us in the present chapter, it is useful to think of heat transfer in microscopic terms (and therefore outside the realm of thermodynamics) as increasing the energy of the system via the *random* motion of the molecules (see Chapter 8). The conversion of heat into work therefore involves, for example, the increase in the potential energy of a piston (*ordered* motion) at the expense of the random motion of the molecules. It is therefore not surprising that heat cannot be converted to work in a cyclic process with an efficiency of unity (eqn 3.50).

In the early nineteenth century however, heat was considered to consist of atoms (caloric) and to be quite distinct from other forms of energy. The conservation of caloric therefore required Q to be a function of state and it is then possible to derive for example the adiabatic equation for a perfect gas $PV^\gamma = \text{constant}$ (eqn 3.46).

Carnot (1796–1831) used the caloric theory in 1824 to produce his two theorems on the efficiency of heat engines which were in time recognised to be of fundamental importance, and it was not until 1850 that Clausius (1822–88) and William Thomson (later Lord Kelvin) (1824–1907) realised that in fact the Carnot theorems did not depend upon the caloric theory. The very careful measurement of the conversion of many different kinds of work into heat by Joule (1818–89) in the period 1843–8 finally established heat as a form of energy transfer.

Clausius first identified the internal energy and the *entropy* (rather than quantity of heat) as functions of state and wrote the equation $dS \geqslant \delta Q/T$. With Kelvin he seems to have been the first to state explicitly that *two* laws of thermodynamics were required. Carathéodory gave the second law of thermodynamics in the form used in Section 3.1 in 1909.

Two interesting accounts of the early work in thermodynamics are: Mendoza (1961) and Klein (1974).

3.9 Conclusions

The establishment of the second law of thermodynamics and of the entropy as a function of state has enormously increased the range of problems that can be treated by thermodynamics beyond those that could be handled using only the first law. The establishment of the entropy as a function of state was in itself sufficient to lead to the derivation of the energy equation (eqn 3.14) and a consideration of the increase of entropy in irreversible processes then led to the correct prediction of the direction of heat flow and the maximum efficiency of a heat engine under given conditions.

The entropy function is not however a very convenient function to work with. The statement, 'in a system within adiabatic walls the entropy cannot decrease', may be applied to any system by enclosing it within a suitable space but most experimental science is carried out at constant temperature and pressure. The entropy statement then requires the system of thermal reservoir (defining the temperature) and experimental system to be treated as a whole. An alternative function of state which concerned only the experimental system at temperature T would clearly be more directly useful than the entropy. Functions of this type (called thermodynamic potentials) are discussed in the next chapter.

3.10 Appendix. The traditional approach to the second law

The efficiency of a heat engine working in a cycle between a hot thermal reservoir at temperature T_A and a cold reservoir at T_B was seen in Section 2.7 to be given by the equation

$$\eta = \frac{-W}{Q_A} = \frac{Q_A - Q_B}{Q_A} \tag{3.55}$$

where $-W$ is the work done by the engine, Q_A the magnitude of the heat extracted from the hot reservoir and Q_B the magnitude of the heat rejected to the cold reservoir, Fig. 3.11. A real engine always rejects heat to the cold reservoir, try touching the exhaust pipe of a motor car, and consequently has an efficiency of less than unity, but this is not required by the first law of thermodynamics. Carnot proposed that even an ideal heat engine would have an efficiency of less than unity and this concept became the basis of the second law of thermodynamics.

Carnot's two theorems may be written:

(1) all reversible engines working between two given heat sources have the same efficiency, independent of the nature of the working substance;
(2) a reversible engine is always more efficient than an irreversible engine working between the same two heat sources.

It is possible to derive these two theorems by considering the effect of coupling engines together but we prefer to derive them in Section 3.7 after introducing the concept of entropy.

There is only one cycle which can possibly be reversible if only two heat sources (at temperature T_A and T_B) are available because heat cannot be

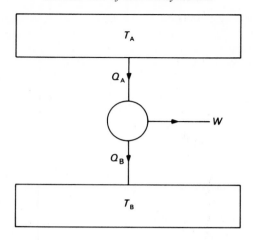

Fig. 3.11. An engine absorbs heat from one thermal reservoir, does work, rejects heat to a second thermal reservoir at a lower temperature and then returns to its original state. The whole operation is called a cycle.

transferred reversibly across a finite temperature difference. The Carnot cycle therefore consists of two isothermals (at T_A and T_B) joined by two adiabats. The shape of the Carnot cycle on a P–V diagram will depend upon the working substance and is shown in Fig. 3.12 for a perfect gas. A direct calculation of the efficiency of an engine following the reversible cycle shown in Fig. 3.12 leads to (Exercise 3.3)

$$\eta = \frac{Q_A - Q_B}{Q_A} = \frac{T_A - T_B}{T_A} \qquad (3.56)$$

and according to Carnot this is the maximum possible efficiency.

Note that, since from Carnot's first theorem the efficiency of the engine is *independent* of the working substance, eqn 3.56 enables us to define a thermodynamic scale of temperature by

$$\frac{Q_A}{Q_B} = \frac{T_A}{T_B} \qquad (3.57)$$

and a choice of a number for one reproducible fixed point. The thermodynamic scale is equivalent to the perfect gas scale but can be determined experimentally even at temperatures near absolute zero where the perfect gas law does not hold.

The efficiency of a heat engine is from eqn 3.56 less than unity unless T_B is zero. It will be shown in Section 4.4 that 0 K is not an experimentally attainable temperature so η is always less than unity.

The Carnot theorems therefore place a restraint on the efficiency of a heat

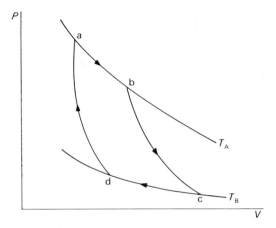

Fig. 3.12 The only reversible cycle for an engine working between two infinite heat reservoirs at T_A and T_B consists of a reversible isothermal expansion at T_A, a reversible adiabatic expansion to temperature T_B, an isothermal contraction at T_B and a final adiabatic contraction. In this way no heat is transferred across a finite temperature difference. The shape of the P–V diagram will depend upon the working substance, here a perfect gas, but the efficiency of the engine is independent of the working substance.

engine which is not required by the first law of thermodynamics. This restraint became the basis of a new law, the second law of thermodynamics, in the nineteenth century and may be expressed in several equivalent ways. The most popular are:

1 Kelvin–Planck: it is impossible to construct an engine working in a cycle which will extract heat Q_A from a hot reservoir and perform work equivalent to Q_A i.e. a cyclic heat engine requires both a source and a sink of heat which must be at different temperatures.

2 Clausius: consider the engine shown in Fig. 3.11 with the arrows reversed. Heat is taken from a cold body to a hot body, a refrigerator, while work is performed on the engine. It is impossible to construct a refrigerator working in a cycle which will extract Q_B from the cold reservoir and deliver Q_A to the hot reservoir without the aid of work W on the engine.

These two statements can be shown to be equivalent by considering an engine and a refrigerator coupled together.

Clausius now took the vital step which enables us to write the second law of thermodynamics in terms of a new function of state. Consider any

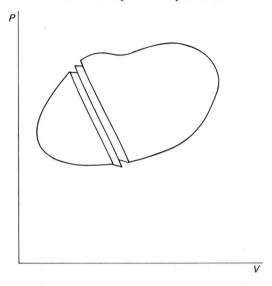

Fig. 3.13　An arbitrary reversible cycle may be considered to be made up of a series of small Carnot cycles.

reversible cycle to be divided into a series of Carnot cycles of infinitesimal area, Fig. 3.13. In each infinitesimal cycle the efficiency is given by eqn 3.56 and, reverting to the sign convention that heat added to a system is taken to be positive,

$$\frac{dQ_A}{T_A} + \frac{dQ_B}{T_B} = 0 \tag{3.58}$$

The summation over the whole reversible cycle is therefore

$$\oint \frac{dQ_R}{T} = 0 \tag{3.59}$$

but this was just the condition required for a function of state so we may write,

$$dS = \frac{dQ_R}{T} \tag{3.60}$$

as the definition of a new function of state called the entropy.

Equation 3.60 is derived by a completely different route in Section 3.1 and the entropy function developed in following sections of Chapter 3.

Exercises

3.1. Using the equations of state given in Exercise 2.4 find expressions for (a) the internal energy and (b) $C_P - C_V$. (c) Show that C_V is independent of the volume for all the equations of state. (d) Find the equations for a reversible adiabatic change.

3.2. Find expressions for the changes of entropy of the system in Exercises 2.5–2.8.

3.3 Verify eqn 3.50 by using (a) a perfect monatomic gas and (b) a van der Waals gas as the working substance in a Carnot cycle.

3.4. An engine contains one mole of a monatomic perfect gas. Show that the efficiency of the engine is the same if it is worked in either a Stirling or a Carnot cycle between the same maximum and minimum pressures, temperatures and volumes but that the work done in the Stirling cycle is greater than that in the Carnot cycle.

3.5 Show that the area enclosed on an entropy–temperature diagram by a reversible engine working in a cycle is equal to the net heat absorbed.

3.6. Construct a P–V diagram for a Carnot cycle if the working substance has the equation of state $P = bT^4 = u/3$ where $U = uV$. (Show that the equation for an isentropic change is $T^3 V = \text{constant}$.) Show that eqn 3.49 still applies. (The equations are those for black body radiation. See Section 10.2.)

3.7. (a) A block of constant heat capacity $500 \, \mathrm{J \, K^{-1}}$ is heated to 600 K and then placed in a lake at 300 K. Find the change in the entropy of the system of block plus lake. (b) The block is reheated to 600 K, placed in boiling water and finally replaced in the lake. Find the new change of entropy and explain how the change could be reduced to zero.

3.8. Write down dS when S is expressed as a function of P and V and show that

$$T dS = \frac{C_v}{\alpha V}(V \kappa_T dP + \gamma \, dV)$$

3.9. Show that the entropy increases when two bodies, initially at a small temperature difference Δ, are placed in contact, see eqn 3.36.

3.10. The isothermal compressibility of a body was defined in eqn 3.22. Show that the adiabatic compressibility is given by

$$\kappa_T = \gamma \kappa_S = -\gamma (\partial \ln V / \partial P)_S$$

Where $\gamma = C_P/C_V$.

3.11. The general expression for the velocity of sound c_s in a fluid is

$$c_s^2 = \left(\frac{\partial P}{\partial \rho}\right)_s.$$

Use the result of Exercise 3.10 to show that

$$c_s^2 = (\rho \kappa_s)^{-1} = \gamma / \rho \kappa_T$$

and show that for a perfect gas this expression is equivalent to that given in Exercise 2.16.

3.12. In many real processes it is found that the volume and pressure of a gas change such that $PV^n = $ constant, a polytropic process. (a) What is the value of n for a process which occurs at (i) constant pressure, (ii) constant volume, (iii) for a perfect gas at constant temperature? (b) Show that the temperature of a perfect gas decreases on expansion if $n > 1$. (c) Show that when a perfect gas expands from (P_1, V_1, T_1) to (P_2, V_2, T_2) and $T_1 \neq T_2$:
(i) $\Delta U = C_v(T_2 - T_1)$; (ii) $\Delta W = R(T_2 - T_1)/(n-1) = (P_2 V_2 - P_1 V_1)/(n-1)$;
(iii) $\Delta Q = [C_v - R/(n-1)](T_2 - T_1)$; and (iv) $\Delta S = [C_v - R/(n-1)] \ln (T_2/T_1)$.
(d) Check the above results for a process which occurs at constant pressure or constant volume. (e) Show that $\Delta S = 0$ is equivalent to $n = C_p/C_v$ as found for a reversible adiabatic change in Section 3.6. (f) Show that the heat capacity for a polytropic process may be written

$$C_v + T\left(\frac{\partial P}{\partial T}\right)_V \left(\frac{\partial V}{\partial T}\right)_{PV^n}$$

and evaluate this for a perfect gas when $n = 2$.

3.13. The reversible Otto cycle, an idealization of the petrol internal combustion engine, has a lower efficiency than a Carnot engine working between the same maximum and minimum temperatures but is a closer approximation to a workable cycle. The cycle consists of 4 parts:

ab   Isentropic compression from (V_a, T_1) to (V_b, T_2), where V_a/V_b is called the compression ratio r.

bc   Heating at constant volume from T_2 to T_3.

cd   Isentropic expansion to (V_a, T_4).

da   Cooling at constant volume to (V_a, T_1).

Draw this cycle and show that for 1 mole of a perfect gas the magnitude of the work done by the gas is

$$|W| = C_v(T_3 - T_2 + T_1 - T_4)$$
$$\eta = 1 - (T_4 - T_1)/(T_3 - T_2) = 1 - (T_4/T_3) = 1 - r^{1-\gamma}$$

Why is the efficiency of a Stirling cycle the same as that of a Carnot engine working between the same thermal reservoirs while the Otto cycle has a lower efficiency?

3.14. The compression ratio, r, in Exercise 3.13 can not be greater than about 10 in a petrol engine because the petrol vapour explodes spontaneously at higher temperatures. What is the efficiency of an Otto engine using air, $C_v = 5R/2$, as a working substance and how does it compare with a Carnot engine working between the same maximum and minimum temperatures of 1500 K and 300 K?

3.15. A practical heat pump might operate between 5°C and 60°C and be driven by a machine with efficiency 0.3 that of a Carnot cycle. What is the maximum ratio of heat delivered to work supplied for this system?

3.16. A refrigerator works between 30°C and −15°C. It extracts 950 kJ of heat for each 190 kJ of work supplied. What is its coefficient of performance and what fraction is this of the theoretical maximum value?

4

Further concepts of thermodynamics

In Chapter 3 the development of the second law of thermodynamics and of the entropy as a function of state led to the derivation of equations such as 3.24 which relate measurable parameters of any system. Thermodynamics is not a complete self-contained description of the thermal properties however since to obtain the *individual* quantities, such as the coefficient of thermal expansion of the system, a certain amount of information has to be provided either from experiment or else from microscopic theory (statistical mechanics, see Chapter 7). The *minimum* amount of information which is required before all the parameters of a system can be calculated using the laws of thermodynamics is therefore of some importance.

In a simple system only two parameters are required to define the state of the system and five variables (T, S, U, P, V) have so far been considered. An equation of the form $f(x, y)$, where x and y are chosen from the above five variables, from which all the thermodynamic properties of the system can be calculated *without performing an integration* is called a *fundamental equation*. As has already been stressed this equation must be obtained from experimental or statistical mechanics. Integration is not an acceptable operation because it would introduce an arbitrary constant into the thermodynamic quantities.

4.1 The fundamental equations

The fundamental equation for a simple system may be written in a number of ways depending upon the choice of the variables x and y. The internal energy expressed as a function of entropy and volume, $U(S, V)$ is a fundamental equation. The quantities S and V are called *natural variables* for U. A knowledge of $U(x, y)$ where x and y are not equal to S and V is not a fundamental equation since some thermodynamic quantities will then be defined only to within a constant of integration.

As a simple example suppose that the equation for the internal energy of a monatomic perfect gas as a function of T and V

$$U = \tfrac{3}{2} N k T$$

has been obtained. Then $C_V = 3/2\,Nk$ so

$$S = \int \frac{C_V}{T}\,\mathrm{d}T = \tfrac{3}{2}\,Nk\ln T + \text{constant}$$

and S is defined only to within the integration constant. If $U(S, V)$ is known then the equation of state and all other properties can be calculated without performing an integration (Exercise 4.1). Similarly $S(U, V)$ is a fundamental equation.

The two fundamental equations $U(S, V)$ and $S(U, V)$ have natural variables which are difficult to evaluate by experiment. The purely mathematical problem is to transform, say $U(S, V)$ to a new fundamental equation with more convenient natural variables. A change of variables of this type is called a Legendre transformation and the new function is called a thermodynamic potential.

The most important thermodynamic potentials for simple systems are tabulated in Table 4.1 along with their natural variables and differential form. The Helmholtz free energy (F) is so called because for a system at constant temperature, that is to say in thermal contact with a heat reservoir at temperature T, the reversible work done in a change of volume from V_A to V_B is simply

$$-\int_{V_A}^{V_B} P\,\mathrm{d}V = F(\mathrm{B}) - F(\mathrm{A})$$

That is, the work done by the gas in a reversible isothermal process is equal to the decrease in the Helmholtz free energy of the system.

Notice that $\mathrm{d}F$ is zero (F is constant) for a reversible process at constant volume and temperature. The Helmholtz free energy $F(T, V)$ is perhaps the most important of all the thermodynamic potentials because it is the quantity which can be calculated most directly by the methods of statistical mechanics (see Chapter 7).

Table 4.1. The definitions and form of the differentials of the enthalpy and Gibbs and Helmholtz free energies

Name	Symbol	Natural Variables	Definition	Differential
Helmholtz free energy	F	(T, V)	$U - TS$	$\mathrm{d}U - T\,\mathrm{d}S - S\,\mathrm{d}T =$ $-P\,\mathrm{d}V - S\,\mathrm{d}T$
Gibbs free energy	G	(T, P)	$F + PV$	$\mathrm{d}F + P\,\mathrm{d}V + V\,\mathrm{d}P =$ $-S\,\mathrm{d}T + V\,\mathrm{d}P$
Enthalpy	H	(S, P)	$U + PV$	$\mathrm{d}U + P\,\mathrm{d}V + V\,\mathrm{d}P =$ $T\,\mathrm{d}S + V\,\mathrm{d}P$

Some examples will now be given of the kind of manipulation of a fundamental equation required to obtain other thermodynamic quantities. The definitions

$$F = U - TS \tag{4.1}$$
$$dF = -P\,dV - S\,dT \tag{4.2}$$
$$= \left(\frac{\partial F}{\partial V}\right)_T dV + \left(\frac{\partial F}{\partial T}\right)_V dT \tag{4.3}$$

lead immediately to the identities

$$\left(\frac{\partial F}{\partial V}\right)_T = -P \tag{4.4}$$

$$\left(\frac{\partial F}{\partial T}\right)_V = -S \tag{4.5}$$

$$U = F + TS = F - T\left(\frac{\partial F}{\partial T}\right)_V = -T^2\left[\frac{\partial(F/T)}{\partial T}\right]_V. \tag{4.6}$$

Equation 4.6 is often called the Gibbs-Helmholtz equation. Notice that P, S and U are all expressed in terms of F or of differentials of F as required if $F(T, V)$ is to be a fundamental equation.

The specific heat at constant volume also follows immediately from

$$C_V = T\left(\frac{\partial S}{\partial T}\right)_V = \left(\frac{\partial U}{\partial T}\right)_V = -T\left(\frac{\partial^2 F}{\partial T^2}\right)_V. \tag{4.7}$$

Some other relations may be found in Exercise 4.2.

The fact that F is a function of state may be applied (in exactly the same way as in eqn 3.12) to eqn 4.2 to obtain the relation

$$\left(\frac{\partial P}{\partial T}\right)_V = \left(\frac{\partial S}{\partial V}\right)_T. \tag{4.8}$$

In fact eqn 4.8 was obtained by a more indirect method in eqn 3.18.

The Helmholtz free energy has been discussed in some detail because of its great importance in theoretical physics. The Gibbs free energy with natural variables (T, P) is of most importance in chemistry since

$$dG = V\,dP - S\,dT \tag{4.9}$$

and therefore a reversible process occurring at constant temperature and pressure has dG equal to zero and G constant. The most important application of the Gibbs free energy in physics is to changes of phase such as the solid–liquid transition (Section 5.4).

The enthalpy is of greatest use in chemical engineering because of its importance in flow processes such as the throttling process discussed in

Section 5.3. The equation

$$dH = T\,dS + V\,dP = d\,Q_R + V\,dP \tag{4.10}$$

shows that for a reversible process at constant pressure the change in the enthalpy is equal to the quantity of heat transferred.

$$C_P = \left(\frac{d\,Q_R}{d\,T}\right)_P = \left(\frac{\partial H}{\partial T}\right)_P \tag{4.11}$$

The fundamental equations have been defined as equations from which all other functions of the system can be calculated without performing an integration. If integration is permitted, that is to say the definition of some terms only to within a constant is acceptable, then only the function of state $f(P, V, T)$ and the heat capacity of the system as a function of temperature at one given pressure or volume are required. If say $C(V_A, T)$ was known from experiment then the entropy could be calculated relative to some reference state (V_A, T_A) using eqn 3.27 and the volume dependence of the specific heat is found from the equation of state using eqn 3.20.

4.2 The Maxwell relations

The equality between $(\partial P/\partial T)_V$ and $(\partial S/\partial V)_T$ derived in eqn 4.8 using the fact that F was a function of state and dF a perfect differential, is one example of a set of equations of great practical importance known as the Maxwell equations. Using eqns 4.8, 4.9, and 4.10 and the second law of thermodynamics in the form $dU = T\,dS - P\,dV$, the four Maxwell relations for simple systems are

$$\left(\frac{\partial P}{\partial T}\right)_V = \left(\frac{\partial S}{\partial V}\right)_T \tag{4.12}$$

$$\left(\frac{\partial V}{\partial T}\right)_P = -\left(\frac{\partial S}{\partial P}\right)_T \tag{4.13}$$

$$\left(\frac{\partial T}{\partial P}\right)_S = \left(\frac{\partial V}{\partial S}\right)_P \tag{4.14}$$

$$\left(\frac{\partial T}{\partial V}\right)_S = -\left(\frac{\partial P}{\partial S}\right)_V. \tag{4.15}$$

The eqns 4.12–4.15 are not of course independent, any of them can be manipulated by using the relations of functions of two variables (Appendix I) to give the other three. The importance of eqn 4.12 was discussed in Section 3.2. The left-hand side of eqn 4.13 may also be calculated directly from the equation of state and be used to replace the quantity $-(\partial S/\partial P)_T$ in any equation in which it occurs.

4.3 Thermodynamic equilibrium

The second law of thermodynamics was shown in Chapter 3 to lead to the conclusion that the entropy of a system within adiabatic walls could not decrease. In thermal equilibrium the entropy of the system is constant at its maximum value for a given internal energy and volume. Consider two subsystems of constant volume within the adiabatic enclosure joined by a diathermal wall. Then

$$U = U_1 + U_2 \quad \mathrm{d}U = \mathrm{d}U_1 + \mathrm{d}U_2 = 0 \tag{4.16}$$

since energy is conserved, and after thermal equilibrium has been established between the two subsystems the entropy is a maximum, so

$$\mathrm{d}S = \mathrm{d}S_1 + \mathrm{d}S_2 = 0. \tag{4.17}$$

Now

$$\mathrm{d}S = \left(\frac{\partial S_1}{\partial U_1}\right)_{V_1} \mathrm{d}U_1 + \left(\frac{\partial S}{\partial U_2}\right)_{V_2} \mathrm{d}U_2$$

$$= \left(\frac{1}{T_1} - \frac{1}{T_2}\right) \mathrm{d}U_1$$

using the definition of the absolute temperature from the second law of thermodynamics. The equilibrium condition (eqn 4.17) therefore requires that $T_1 = T_2$ for thermal equilibrium between the subsystems. In this sense the zeroth law of thermodynamics (Section 2.1) is not a necessary postulate for the construction of thermodynamics, it is contained within the second law. The equilibrium state has not been shown to be stable since only the fact that $\mathrm{d}S = 0$ has been used so far. The maximum condition on the entropy also requires $\mathrm{d}^2 S < 0$ but this will not be proved here.

The equilibrium condition for a system in contact with a thermal reservoir at temperature T_R can also be written in terms of the entropy maximum of the combination of system and reservoir enclosed within adiabatic walls (Fig. 4.1) but it is obviously more convenient to express the equilibrium condition in terms only of the parameters of the system of interest and the temperature of the reservoir (T_R). The enclosed combination of thermal reservoir and system of interest has, by analogy with eqns 4.16 and 4.17

$$\mathrm{d}(U + U_R) = 0 \quad \mathrm{d}(S + S_R) = 0 \tag{4.18}$$

subject now also to the condition that the temperature always be T_R. Now at *constant volume* $\mathrm{d}U_R = T_R \, \mathrm{d}S_R$ and therefore using eqn 4.18

$$\mathrm{d}U + T_R \, \mathrm{d}S_R = \mathrm{d}U - T_R \, \mathrm{d}S = 0.$$

Therefore

$$\mathrm{d}(U - T_R S) = 0 \tag{4.19}$$

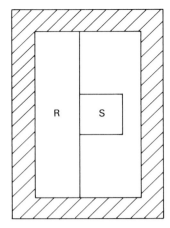
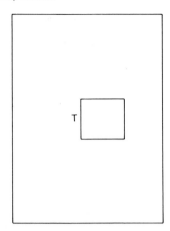

Fig. 4.1 A system (S) of fixed volume in thermal equilibrium at temperature T may be treated either by considering it as attached to a thermal reservoir (R) within adiabatic walls, (the total internal energy is then constant and the entropy a maximum) or as a system with defined temperature T and volume V for which the Helmholtz free energy is a minimum.

since T_R is constant. The condition for thermal equilibrium for a system at constant volume in contact with a thermal reservoir is therefore given by eqn 4.19, which is equivalent to $dF = 0$ where $F = U - TS$ is the Helmholtz free energy. A consideration of d^2F shows that F must be a minimum. Similarly for a system in contact with a thermal reservoir and maintained at constant pressure the Gibbs free energy is a minimum (Exercise 4.4).

An example may clarify the concept of the Helmholtz free energy being a minimum subject to the conditions that T and V be constant. In Exercise 6.6 it is shown that a perfect crystal, a regular array of atoms, cannot exist at temperatures above absolute zero. If n atoms are removed from their sites, involving an energy expenditure of $n\varepsilon$ then the Helmholtz free energy is

$$F = n\varepsilon - kT\left[N\ln N - (N-n)\ln(N-n) - n\ln n\right] \qquad (4.20)$$

where N is the total number of sites in the perfect crystal. The number of defects (n) of a crystal in thermal equilibrium at temperature T is now found by setting $(\partial F / \partial n)_T$ equal to zero in eqn 4.20.

4.4 Third law of thermodynamics

The first and second laws of thermodynamics which have been developed in earlier sections are sufficient to show that the thermodynamic temperature can never be negative (once the temperature at the triple point of water was defined to be positive) although it is possible to describe the behaviour of certain subsystems by a negative temperature under special conditions (Section 7.8). The lowest thermodynamic temperature is therefore 0 K. The entropy change of a system at constant volume cooled from T_A to T_B, given by

$$\Delta S = \int_{T_A}^{T_B} \frac{C_V}{T} \, dT$$

would therefore become infinite if C_V were independent of temperature (as was predicted by classical physics) and T_B were set equal to zero. The experimental measurements of the specific heats of metals at low temperatures however are of the form

$$C_V = AT^3 + \gamma T \tag{4.21}$$

where A and γ are constants (Fig. 4.2). (See also Section 10.3.) Then $(C_V/T) \to \gamma$ as $T \to _{0\,K}$ and the integral remains finite.

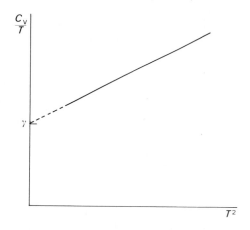

Fig. 4.2   The specific heat of a metal (divided by the temperature) at low temperature as a function of T^2. The straight line has the form of eqn 4.21. The intercept represents the contribution of 'free' electrons (Section 10.1) and is not present for non-metals. See also Figure 10.9

The third law of thermodynamics was originally developed by Nernst for chemical changes and like the second law has been expressed in a number of different forms. The most useful perhaps from the point of view of physics is: 'as the temperature approaches zero the entropy of any system becomes independent of the parameters of the system'.

$$T \to _{0K} \quad S \to s_0 \tag{4.22}$$

where S_0 is a constant. Since the thermodynamic temperature is equal to

$(\partial U/\partial S)_V$ this is equivalent to

$$\left(\frac{\partial U}{\partial S}\right)_{V\to 0\,K} S\to s_0.$$

Since S_0 is a constant, the differentials of the entropy tend to zero as $T_{\to 0\,K}$. Now using eqns 4.13 and 3.23

$$\left(\frac{\partial S}{\partial P}\right)_T = -\left(\frac{\partial V}{\partial T}\right)_P = -\alpha V. \tag{4.23}$$

The coefficient of thermal expansion therfore goes to zero as $T_{\to 0\,K}$ (Fig. 4.3). Similarly using eqn 4.12

$$\left(\frac{\partial S}{\partial V}\right)_T = \left(\frac{\partial P}{\partial T}\right)_V \tag{4.24}$$

goes to zero (Fig. 4.4). The isothermal compressibility, $-1/V(\partial V/\partial P)_T$ however goes to a constant value not to zero.

All specific heats certainly go to zero since

$$C_x = T\left(\frac{\partial S}{\partial T}\right)_x. \tag{4.25}$$

and $(\partial S/\partial T)_x$ must remain finite as $T_{\to 0\,K}$. According to the experimental eqn (4.21) however, for a metal

$$\left(\frac{\partial S}{\partial T}\right)_V = \frac{C_V}{T} \tag{4.26}$$

approaches the constant value γ rather than zero as $T_{\to 0\,K}$.

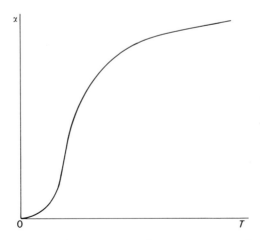

Fig. 4.3 The coefficient of volume expansion goes to zero at low temperature.

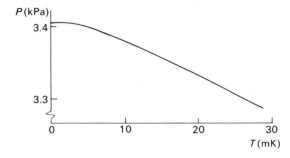

Fig. 4.4 $(\partial P/\partial T)_V$ goes to zero at low temperature as shown here for ^{3}He. See also Section 10.4. After Huiskamp and Lounasmaa (1973)

In fact on the basis of statistical mechanics (Section 7.8), $(\partial S/\partial T)_V$ does approach zero because eqn 4.21 is incorrect at temperatures extremely close to absolute zero ($\approx 10^{-10}$ K) which cannot be reached by experiment. The specific heat then decreases exponentially to zero and $(C_V/T)_{\to 0}$ as $T_{\to 0\,\mathrm{K}}$. The temperatures from which it is safe to extrapolate measurements to absolute zero must therefore always be carefully considered.

A classical perfect gas cannot exist at low temperatures since the specific heat of the gas is independent of the temperature and the entropy would diverge as the temperature goes to zero (eqn 3.28). The quantum–mechanical gases discussed in Section 10.1 could however in principle continue to exist to the lowest temperatures. In fact, apart from atomic hydrogen, only the two isotopes of helium remain even as liquids (at atmospheric pressure) at the lowest temperatures attainable by experiment (see Sections 10.1 and 10.4).

The first and second laws of thermodynamics have both been expressed in earlier sections in terms of the impossibility of constructing a perpetual motion machine. The third law can also be written in a restrictive manner: 'it is not possible to cool a substance to the temperature $T=0$ K.' The distinction is to be made between *approaching* 0 K as closely as desired and actually attaining the mathematical value of zero. This statement is consistent with the earlier one that $S_{\to S_0}$ as will now be shown. The most efficient method of cooling the system is a reversible adiabatic expansion for which $\mathrm{d}S = 0$. Then from eqn 3.15

$$T\,\mathrm{d}S = C_V\,\mathrm{d}T + T\left(\frac{\partial P}{\partial T}\right)_V \mathrm{d}V = 0. \qquad (4.27)$$

The temperature drop in a reversible adiabatic expansion is therefore

$$\Delta T = -\frac{T}{C_V}\left(\frac{\partial P}{\partial T}\right)_V \Delta V. \qquad (4.28)$$

Now since $(\partial P/\partial T)_V$ goes to zero as $T \rightarrow 0\,\mathrm{K}$ (eqn 4.24) because $S \rightarrow S_0$ the temperature $T = 0\,\mathrm{K}$ cannot be arrived at from any finite temperature T.

The value of S_0 has been left undefined so far and cannot be evaluated on the basis of thermodynamics. In Chapter 6 it is shown that the entropy of a system in true thermodynamic equilibrium does appear to approach zero as $T \rightarrow 0\,\mathrm{K}$. This statement is however only useful for pure crystals since diffusion processes in solids are so slow that an alloy will always be in a metastable state at low temperature. Similarly, systems which form glasses or amorphous solids rather than perfect crystals will have $S \rightarrow S_0$ as $T \rightarrow 0\,\mathrm{K}$ but S_0 is not equal to zero. Thermodynamics leads only to values for the *difference* of the entropy of a system in two given equilibrium states since the entropy was originally introduced in the form of the differential $dS = dQ_R/T$.

*4.5 Appendix. Thermodynamic systems with a variable number of particles

It has been assumed throughout the discussion to this point that the number of particles in a thermodynamic system remains constant as the system goes from an initial equilibrium state (T_A, V_A) to a final state (T_B, V_B) but it is clear that this cannot be true in general. Consider, for example, 1 mole of hydrogen gas containing N_A diatomic molecules of hydrogen. As the temperature is raised the molecules will dissociate and the atoms ionize until there are $2N_A$ protons and $2N_A$ electrons in the system. The equilibrium number of particles will depend upon the temperature, which increases the number of atoms and molecules in excited states, and upon the volume per particle because once an atom has ionized the electron has to meet a proton in order to recombine. There is therefore a dynamic equilibrium between the different species of particles. A similar problem arises if the temperature of the gas is lowered. At some point a liquid phase will form in contact with the vapour and the number of particles in each phase will be a function of the temperature and pressure of the system. The number of particles could also change due to chemical reactions, which will not be considered here, or if the system of interest is free to exchange particles with a reservoir.

The general expression for the second law of thermodynamics may be written from eqn 2.47

$$T\,dS = dQ_R = \left(\frac{\partial U}{\partial T}\right)_x dT + \sum_{r=1}^{n-1} \left[\left(\frac{\partial U}{\partial x_r}\right)_c - X_r\right] dx_r$$

where the subscript x means that all the $(n-1)x_r$ are held constant, and the subscript c means the temperature and all x other than x_r are held constant, while performing the differentiation. Each term $X_r dx_r$ represents a work term such as $-P\,dV$ or $\mathscr{F}\,dl$.

---

* This section may be omitted on a first reading.

In order to allow for the variation of the number of particles, N, with the thermodynamic state of the system, the energy is now written as $U(S, V, N)$ for example, rather than simply $U(S, V)$,

$$dU = \left(\frac{\partial U}{\partial S}\right)_{V,N} dS + \left(\frac{\partial U}{\partial V}\right)_{S,N} dV + \left(\frac{\partial U}{\partial N}\right)_{S,V} dN$$

$$= T\,dS - P\,dV + \mu\,dN$$

where a new function, the chemical potential, has been defined by,

$$\mu = \left(\frac{\partial U}{\partial N}\right)_{S,V}$$

The definition of the Helmholtz and Gibbs free energy is unchanged, $F = U - TS$, $G = F + PV$, but their differentials become,

$$dF = -S\,dT - P\,dV + \mu\,dN \tag{4.29}$$

$$dG = -S\,dT + V\,dP + \mu\,dN \tag{4.30}$$

and a new set of Maxwell relations may now be derived including N as a variable.

The chemical potential may be written

$$\mu = \left(\frac{\partial U}{\partial N}\right)_{S,V} = \left(\frac{\partial F}{\partial N}\right)_{T,V} = \left(\frac{\partial G}{\partial N}\right)_{T,P}. \tag{4.31}$$

The natural variables for G were seen in Section 4.1 to be the intensive quantities P and T so for a system of identical particles G is simply proportional to N and $\mu(T, P)$ is identical to the Gibbs free energy per particle,

$$G(T, P, N) = N\mu(T, P). \tag{4.32}$$

A new thermodynamic potential function, the grand potential, can now be introduced with natural variables (T, V, μ) since,

$$dG = \mu\,dN + N\,d\mu$$

$$dF = -S\,dT - P\,dV + \mu\,dN$$

$$\therefore d\Omega = d(F - G) = -S\,dT - P\,dV - N\,d\mu. \tag{4.33}$$

The original definition of F and G leads to the definition of the grand potential as

$$\Omega = U - TS - \mu N = -PV. \tag{4.34}$$

The number of particles in the system is given by

$$N = -\left(\frac{\partial \Omega}{\partial \mu}\right)_{T,V} = V\left(\frac{\partial P}{\partial \mu}\right)_{T,V}. \tag{4.35}$$

What is the physical meaning of the chemical potential? When two closed sub-systems are placed in thermal contact it was seen in Section 4.3 that they evolve until they come to a common temperature. Using the same technique, Exercise 4.15, for two sub-systems free to exchange particles it is found that at equilibrium,

$$\frac{\mu_1}{T_1} = \frac{\mu_2}{T_2}. \tag{4.36}$$

Since $T_1 = T_2$ it follows that in thermal equilibrium the chemical potential must be constant throughout the whole system, i.e., particles will diffuse between the two sub-systems until $\mu_1 = \mu_2$.

The chemical potential of a system in a potential gradient may be written as the sum of internal and external contributions,

$$\mu = \mu_i(P, T) + V(\mathbf{r}). \tag{4.37}$$

In a gravitational field for example the external potential for a molecule of mass m at height z is simply mgz. Since μ must be constant throughout the system it follows that μ_i is a function of z, Exercise 4.14. The application of equation 4.37 to semiconductor devices is discussed in detail in Kittel and Kroemer (1980).

Exercises

4.1. The fundament equation $U(S, V)$ of a monatomic perfect gas may be written

$$U = \alpha N k \left(\frac{N}{V}\right)^{\frac{2}{3}} e^{\frac{2S}{3Nk}}$$

where α is a constant. Show that the equation of state $PV = RT$, the principle specific heats and the fundamental equation in the form $F(T, V)$ may all be derived from this equation.

4.2. Derive expressions for the isothermal compressibility and the difference $(C_P - C_V)$ in terms of $F(T, V)$.

4.3. Show that $C_P = -T \left(\dfrac{\partial^2 G}{\partial T^2}\right)_P$

4.4. Show that the equilibrium condition for a system at constant temperature and pressure may be written $dG = 0$.

4.5. Write down dS when S is expressed as a function of T and P and use the Maxwell relations to show that

$$T dS = C_p dT - T \left(\frac{\partial V}{\partial T} \right)_P dP$$

Hence show that the entropy of a perfect gas may be written

$$S(P_B, T_B) - S(P_A, T_A) = C_p \ln \left(\frac{T_B}{T_A} \right) - R \ln \left(\frac{P_B}{P_A} \right)$$

Show that this equation is equivalent to eqn. 3.28.

4.6. Show that

$$\left(\frac{\partial H}{\partial P} \right)_T = V(1 - T\alpha)$$

4.7. Show that for a real gas describable by the virial expansion, eqn 2.10,

$$U = \frac{-RT^2}{V} \left[\frac{dB_v}{dT} + \frac{1}{2V} \frac{dC_v}{dT} + \cdots \right] + U_0$$

$$H = \frac{RT}{V} \left[\left(B_v - T \frac{dB_v}{dT} \right) + \frac{1}{V} \left(C - \frac{T}{2} \frac{dC_v}{dT} \right) + \cdots \right] + H_0$$

4.8. Show that the change in the enthalpy of a system with temperature may be found from measurements of the heat capacity at constant pressure.

4.9. Show that the latent heat of a phase transition is equal to the change in the enthalpy per unit mass of the system.

4.10. Show that $F(T, P)$ is not a fundamental equation by establishing that the volume of the system can only be found by performing an integration.

4.11. Show that $G = -V^2 \left\{ \frac{\partial}{\partial V} \left(\frac{F}{V} \right) \right\}_T$

$$H = F - T \left(\frac{\partial F}{\partial T} \right)_V - V \left(\frac{\partial F}{\partial V} \right)_T$$

4.12. Show that the equations for black body radiation, Section 10.2,

$$U = 3PV = bVT^4$$

where b is a constant, satisfy the third law of thermodynamics.

4.13. Derive the Gibbs-Duhem equation,

$$S\mathrm{d}T - V\mathrm{d}P + N\mathrm{d}\mu = 0$$

from the expression for the Gibbs free energy. Hence show that the volume per particle is given by

$$V/N = \left(\frac{\partial \mu}{\partial P}\right)_T$$

4.14. Show that at height z in a gravitational field the internal chemical potential of a molecule of mass m of a gas at uniform temperature T is given by

$$\left(\frac{\partial \mu_i}{\partial P}\right)_T = -mg\left(\frac{\partial z}{\partial P}\right)_T.$$

and therefore

$$\mathrm{d}P = -\rho g \mathrm{d}z$$

as was obtained in Exercise 2.13 by considering mechanical equilibrium.

4.15. Show that

$$\left(\frac{\partial S}{\partial N}\right)_{U,V} = -\frac{\mu}{T}$$

Hence show that if the entropy of two parts of a body which are free to exchange particles is written $S = S_1 + S_2$, with $N = N_1 + N_2$, the condition for S to be a maximum is $\mu_1/T_1 = \mu_2/T_2$.

5

Further applications of thermodynamics

The fundamental equations of thermodynamics have been established in Chapters 2–4. Four important applications of these equations are discussed in this chapter and other applications are considered using both thermodynamics and statistical mechanics in Chapter 10.

5.1 Reduction of measurements to constant volume

Theoretical calculations in solid-state physics are always made for systems at constant volume but experiments on properties such as the specific heat are normally made at constant (atmospheric) pressure. The pressure required to maintain a system at constant volume can be calculated by considering the volume as a function of temperature and pressure

$$dV = \left(\frac{\partial V}{\partial T}\right)_P dT + \left(\frac{\partial V}{\partial P}\right)_T dP.$$

At constant volume the left-hand side is zero and

$$\Delta P = P(T_B) - P(T_A) = -\int_{T_A}^{T_B} \left[\left(\frac{\partial V}{\partial T}\right)_P \middle/ \left(\frac{\partial V}{\partial P}\right)_T\right] dT$$

is the increase in pressure required to keep the volume constant as the temperature increases from T_A to T_B. This equation may be written

$$\Delta P = \int_{T_A}^{T_B} \alpha \kappa_T^{-1} dT \tag{5.1}$$

using eqns 3.22 and 3.23. When the system is a perfect gas, eqn 5.1 becomes simply

$$\Delta P = \frac{R}{V}(T_B - T_A) = \frac{P\Delta T}{T} \tag{5.2}$$

so the fractional increase in the pressure to hold the system at constant volume is simply equal to the fractional change in the thermodynamic temperature.

The pressure required to hold a solid at constant volume depends upon the temperature of the system since the coefficient of thermal expansion goes to zero at absolute zero (eqn 4.23) but is usually only weakly dependent on temperature near room temperature. The isothermal compressibility of a solid is constant at low temperatures but increases slowly with temperature. At room temperature therefore eqn 5.1 becomes, for a small change in temperature ΔT

$$\Delta P = \overline{\alpha \kappa_T^{-1}} \Delta T \tag{5.3}$$

The values $\alpha = 50 \times 10^{-6} \, \text{K}^{-1}$, $\kappa_T^{-1} = 14 \times 10^{10} \, \text{N m}^{-2}$ for copper at room temperature are typical values for solids. Then a 1 K rise in temperature requires the pressure to change by $7 \times 10^6 \, \text{N m}^{-2}$ (roughly 70 atmospheres) to hold the system at constant volume. The system would therefore have to be enclosed within a high pressure press if direct measurements were to be made at constant volume but by using the laws of thermodynamics a general thermodynamic result will now be obtained relating measurements at constant pressure to those at constant volume.

Suppose that some equilibrium property of a system Φ has been measured as a function of temperature at some pressure P_0. The function $\Phi(T, P_0)$ is therefore known and the function $\Phi(T, V)$ may be related to it by the usual expressions

$$d\Phi = \left(\frac{\partial \Phi}{\partial T}\right)_V dT + \left(\frac{\partial \Phi}{\partial V}\right)_T dV$$

$$\left(\frac{\partial \Phi}{\partial T}\right)_V = \left(\frac{\partial \Phi}{\partial T}\right)_P - \left(\frac{\partial \Phi}{\partial V}\right)_T \left(\frac{\partial V}{\partial T}\right)_P. \tag{5.4}$$

Then

$$\Phi(V_0, T) - \Phi(V_0, T_0) = \Phi(P_0, T) - \Phi(P_0, T_0) - \int_{T_0}^{T} \left(\frac{\partial \Phi}{\partial V}\right)_T \left(\frac{\partial V}{\partial T}\right)_P dT$$

$$= \Phi(P_0, T) - \Phi(P_0, T_0)$$

$$- \int_{T_0}^{T} \left(\frac{\partial \Phi}{\partial P}\right)_T \left(\frac{\partial P}{\partial V}\right)_T \left(\frac{\partial V}{\partial T}\right)_P dT. \tag{5.5}$$

The second term on each side of eqn 5.5 will cancel if (P_0, V_0, T_0) defines a state of the system. The evaluation of the integral requires the temperature dependence of the isothermal compressibility (usually small for a solid), the coefficient of expansion and the pressure dependence of the quantity to be evaluated. It is not therefore necessary to hold the volume of the system constant but the pressure apparatus must achieve a sufficiently high pressure for $(\partial \Phi / \partial P)_T$ and $(\partial V / \partial P)_T$ to be evaluated. The equation simplifies considerably when only the volume in the integral is found to be a strong function of temperature

$$\Phi(V_0, T) = \Phi(P_0, T) + \overline{\left(\frac{\partial \Phi}{\partial P}\right)_T} \kappa_T^{-1} \ln\left(\frac{V_T}{V_0}\right). \qquad (5.6)$$

The correction to constant volume at a given temperature therefore requires a minimum of four measurements, for two of which high-pressure apparatus is required. The important point is that these quantities can all be evaluated separately, perhaps in different laboratories, and then the final correction made.

5.2 The principal specific heats

The relationship between the specific heats at constant pressure and constant volume is of particular importance since the specific heat is the most easily measurable thermal quantity. The required equation has already been established (eqns 2.34 and 3.14) and can also be obtained directly from eqn 5.4 by taking Φ to be the entropy. Then

$$\left(\frac{\partial S}{\partial T}\right)_V = \left(\frac{\partial S}{\partial T}\right)_P - \left(\frac{\partial S}{\partial V}\right)_T \left(\frac{\partial V}{\partial T}\right)_P$$

or

$$C_P - C_V = T\left(\frac{\partial S}{\partial V}\right)_T \left(\frac{\partial V}{\partial T}\right)_P$$

$$= T\left(\frac{\partial P}{\partial T}\right)_V \left(\frac{\partial V}{\partial T}\right)_P \qquad (5.7)$$

using the definitions of the specific heats and the Maxwell relation (eqn 4.12). The right-hand side of eqn 5.7 may be rewritten using the standard relationship

$$\left(\frac{\partial P}{\partial T}\right)_V \left(\frac{\partial V}{\partial P}\right)_T \left(\frac{\partial T}{\partial V}\right)_P = -1 \qquad (5.8)$$

$$C_P - C_V = -T\left(\frac{\partial V}{\partial T}\right)_P^2 \left(\frac{\partial P}{\partial V}\right)_T \qquad (5.9)$$

$$= \frac{TV\alpha^2}{\kappa_T}. \qquad (5.10)$$

Since $(\partial P/\partial V)_T$ is never positive for a system in thermal equilibrium (see Section 2.2) and $(\partial V/\partial T)_P^2$ cannot be negative because of the squared term, eqn 5.10 establishes the important result that $C_P \geqslant C_V$. The specific heat at constant pressure is equal to that at constant volume if $(\partial V/\partial T)_P$ is zero, as occurs for water near 4°C, and $(C_P - C_V)$ also tends to zero at low temperatures since the coefficient of thermal expansion (α) is zero at $T=0$ K but the isothermal compressibility (κ_T) is not. In fact (C_P-C_V) goes to zero more rapidly than either C_P or C_V separately since (C_V/α) is found to be a constant at low temperatures. The specific heat of an insulator, for example is proportional to T^3 at low temperatures (see Section 10.3) so (C_P-C_V) $\approx T^7$.

The behaviour of the specific heat at the critical point is also of importance. At the critical point

$$\left(\frac{\partial P}{\partial V}\right)_T = 0$$

(Section 2.2) so the isothermal compressibility goes to infinity. Since the right-hand side of eqn 5.7 may be rewritten using eqn 5.8

$$C_P - C_V = -T\left(\frac{\partial P}{\partial T}\right)_V^2 \bigg/ \left(\frac{\partial P}{\partial V}\right)_T \tag{5.11}$$

(C_P-C_V) goes to infinity at the critical point. Experiments suggest that the specific heat at the critical volume also goes to infinity as the temperature approaches T_c but the divergence is much weaker for C_V than for C_P (Section 10.4).

The high isothermal compressibility near the critical point leads to large fluctuations in the density of the system over regions large enough to cause the scattering of a beam of light passing through the system, a phenomenon known as critical opalescence.

The correction of the measured specific heat (C_P) as a function of temperature to the specific heat at constant volume using eqn 5.10 requires the complete temperature dependence of the volume per unit mass, the coefficient of thermal expansion and the isothermal compressibility. This large amount of experimental information is rarely available but fortunately the quantity

$$A = \frac{V\alpha^2}{\kappa_T C_P^2}$$

is found to be nearly independent of temperature (Zemansky and Dittman, 1981). The equation then becomes

$$C_P - C_V = AC_P^2 T \tag{5.12}$$

The values of V, α and κ_T for the system are therefore now required at only *one* temperature. Equation 5.12 is called the Nernst–Lindemann equation (see Exercise 5.1).

A discussion of the specific heat of solids as a function of temperature using statistical mechanics will be found in Section 10.3. The essential features are that both C_P and C_V go to zero at absolute zero and that C_V tends to a constant value at sufficiently high temperature. The specific heat at constant pressure however does not become constant at high temperature, as may be seen from eqn 5.12.

5.3 Cooling and liquefaction of gases

The liquefaction of the so-called permanent gases such as air, nitrogen, or helium, is one of the most important applications of thermodynamics both for industrial uses and for low-temperature research. The essential feature of the gas–liquid transition is that a gas cannot be liquefied by increasing the pressure unless the temperature is below the critical temperature (Section 2.2). The simplest method of liquefying a gas is therefore to place it in contact with a reservoir at a sufficiently low temperature. For example a Stirling engine refrigerator (Section 3.7) using hydrogen or helium as a working substance can run at a sufficiently low temperature (75 K) for air to liquefy at atmospheric pressure.

Alternatively a gas may cool after a free expansion, an adiabatic expansion, or a throttling process, although only the reversible adiabatic expansion invariably leads to a decrease in temperature as will now be shown. The free expansion of a gas was discussed in Section 3.3. The internal energy of the gas is constant in this process

$$dU = \left(\frac{\partial U}{\partial V}\right)_T dV + \left(\frac{\partial U}{\partial T}\right)_V dT = 0.$$

Now from eqns 2.32 and 3.14

$$\left(\frac{\partial U}{\partial T}\right)_V = C_V$$

$$\left(\frac{\partial U}{\partial V}\right)_T = T\left(\frac{\partial P}{\partial T}\right)_V - P.$$

Therefore

$$C_V \, dT + \left[T\left(\frac{\partial P}{\partial T}\right)_V - P\right] dV = 0$$

$$T(B) - T(A) = -\int_{V_A}^{V_B} \left[\frac{T\left(\frac{\partial P}{\partial T}\right)_V - P}{C_V}\right] dV. \qquad (5.13)$$

A free expansion always leads to a cooling of a van der Waals gas (Exercise 5.2) but occurs without change of temperature for a perfect gas as has already been seen in Section 3.3 since $(\partial U/\partial V)_T$ is zero for a perfect gas.

A reversible adiabatic expansion of *any* gas always leads to a decrease in temperature. Since the process occurs at constant entropy the important parameter is $(\partial T/\partial V)_S$. Using the Maxwell-relation equation (4.15)

$$\left(\frac{\partial T}{\partial V}\right)_S = -\left(\frac{\partial P}{\partial S}\right)_V$$

$$= -\left(\frac{\partial T}{\partial S}\right)_V \left(\frac{\partial P}{\partial T}\right)_V$$

$$= \frac{T}{C_V}\left(\frac{\partial P}{\partial V}\right)_T \left(\frac{\partial V}{\partial T}\right)_P \tag{5.14}$$

Since $(\partial P/\partial V)_T < 0$ for equilibrium and $(\partial V/\partial T)_P > 0$ for a gas, $(\partial T/\partial V)_S < 0$. A gas always cools in a reversible adiabatic expansion and the cooling is greater than for a given volume change in a free expansion as may be seen by comparing eqns 5.13 and 5.14.

The temperature of a gas may therefore be reduced by using it as the working substance in a suitable engine provided that the operation is sufficiently close to that of a reversible engine. However at low temperatures friction is always a serious problem because normal engine lubricants cannot be used and only gas lubrication of the engine pistons is possible. The liquefaction of helium is therefore carried out in the Collins liquefier by reducing the temperature to 10 K in an expansion engine and then achieving the final temperature drop to 4 K in the throttling process which will now be described.

The principle of a throttling process is illustrated in Fig. 5.1. A gas at pressure P_A is initially contained within an adiabatic enclosure, one wall of which is formed by a piston behind a porous plug. The two pistons are now moved so that the pressure on the left of the plug is always P_A and on the right of the plug P_B as shown in the figure. The final equilibrium state is

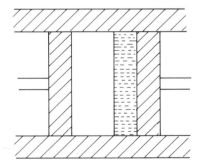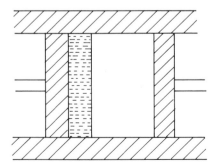

Fig. 5.1 In a throttling process the gas is thermally isolated and the pistons moved so that the pressure is always P_A on the left-hand side of the porous plug and P_B on the right-hand side. The gas passes through non-equilibrium states in the porous plug but thermodynamics may be applied to the initial and final equilibrium states.

(P_B, V_B). The passage of the gas through the porous plug is an irreversible process but thermodynamics may be used to find the relationship between the initial and final equilibrium states.

Since the whole system is thermally isolated, Q is zero and the change in the internal energy of the gas is equal to the net work done on the gas. That is

$$U(B) - U(A) = - \int_{V_A}^{0} P_A \, dV - \int_{0}^{V_B} P_B \, dV$$

$$= P_A V_A - P_B V_B$$

since both changes take place at constant pressure. Therefore

$$U(B) + P_B V_B = U(A) + P_A V_A. \tag{5.15}$$

Since the enthalpy (H) is defined as $U + PV$ (Table 4.1) eqn 5.15 is equivalent to the statement that the initial and final states of a gas in a throttling process are states of equal enthalpy. The intermediate states which the gas passes through in the porous plug are nonequilibrium states to which thermodynamics cannot be applied. The entropy of the gas therefore increases in a throttling process. Consider unit mass of gas passing through the plug. Then

$$T \, ds = du + P \, dv$$

where the lower case symbol is used to indicate unit mass. Now

$$dh = du + P \, dv + v \, dP.$$

Therefore

$$T \, ds = dh - v \, dP. \tag{5.16}$$

The initial and final states of the gas have the same enthalpy, so dh is zero but dP is negative since the pressure decreases. The entropy of unit mass of gas therefore increases in a throttling process by an amount

$$s(B) - s(A) = - \int_{P_A}^{P_B} \frac{v \, dP}{T}. \tag{5.17}$$

The integral is as usual evaluated for any reversible path joining P_B and P_A and may be calculated if the equation of state of the gas is known. A perfect gas has the particularly simple relation

$$s(B) - s(A) = - R \int_{P_A}^{P_B} \frac{dP}{P} = R \ln\left(\frac{P_A}{P_B}\right) > 0. \tag{5.18}$$

The temperature change between initial and final equlibrium states in a throttling process is due to a change of pressure at constant enthalpy, $(\partial T / \partial P)_h$. Using eqn 5.16.

$$\left(\frac{\partial h}{\partial P}\right)_T = T \cdot \left(\frac{\partial s}{\partial P}\right)_T + v.$$

Therefore

$$-\left(\frac{\partial h}{\partial T}\right)_P \left(\frac{\partial T}{\partial P}\right)_h = T\left(\frac{\partial s}{\partial P}\right)_T + v$$

$$\left(\frac{\partial T}{\partial P}\right)_h = \frac{1}{C_P}\left[T\left(\frac{\partial v}{\partial T}\right)_P - v\right]. \tag{5.19}$$

The last equation follows from the identity

$$C_P = T\left(\frac{\partial s}{\partial T}\right)_P = \left(\frac{\partial h}{\partial T}\right)_P \tag{5.20}$$

(eqn 4.11) and the Maxwell relation (eqn 4.13). The quantity $(\partial T/\partial P)_h$ is called the Joule–Kelvin coefficient since they first performed the porous plug experiment. The temperature change after a finite pressure drop may be found by integrating eqn 5.19. After a small pressure drop ΔP, the temperature change is given by

$$\Delta T = \frac{1}{C_P}\left[T\left(\frac{\partial v}{\partial T}\right)_P - v\right]\Delta P. \tag{5.21}$$

Since C_P is positive and ΔP negative (pressure *drop*) the gas cools if the bracket is positive.

The temperature change for a perfect gas is zero but for a van der Waals gas at low density (i.e. ignoring second-order terms)

$$\Delta T = \frac{1}{C_P}\left(\frac{2a}{RT} - b\right)\Delta P. \tag{5.22}$$

At low temperature the bracket is positive and the gas cools, at a temperature equal to $2a/Rb$ there is no change of temperature and at higher temperature the gas warms. The temperature at which there is no temperature change after the throttling process is called the *inversion temperature*. The inversion temperature for the van der Waals gas is related to the critical temperature, using eqn 2.14, and to the Boyle temperature (page 20)

$$T_i = \frac{27}{4} T_c = 2T_B. \tag{5.23}$$

The general equation of the inversion curve is found by setting ΔT equal to zero in eqn 5.21

$$T\left(\frac{\partial v}{\partial T}\right)_P = v. \tag{5.24}$$

In a real gas the inversion temperature (given by eqn 5.24) is a function of the pressure, as shown schematically in Fig. 5.2. The inversion curve is shown as a dotted line. The maximum cooling occurs for a given T_A if the pressure P_A is

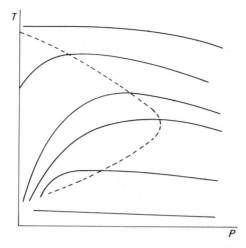

Fig. 5.2 In a real gas the inversion temperature is a function of pressure. The full lines are curves of constant enthalpy, the dotted line the inversion curve. The maximum inversion temperature occurs when the inversion curve reaches the T-axis. Since the initial and final states of a Joule-Kelvin expansion lie on a curve of constant enthalpy it can be seen that the maximum cooling occurs if the initial state lies on the inversion curve.

such that (P_A, T_A) lies on the inversion curve. There is a maximum inversion temperature above which no cooling is possible. The relationship between the maximum inversion temperature and the Boyle temperature for a van der Waals gas (given by eqn 5.23) is also correct to within 10 per cent for real gases.

The change of temperature of a real gas in a Joule–Kelvin expansion depends upon the *sum* of the differences from Boyle's law and Joule's law for a perfect gas (Exercise 5.3). The inversion curve is therefore a stringent test of an equation of state for a real gas. An equation of the form $PV = RT + B_P P + C_P P^2$ is inadequate (Exercise 5.5) at low temperatures because the calculated pressure on the inversion curve goes to infinity instead of showing a maximum as in Fig. 5.2.

The most difficult gas to liquefy is helium, which has a critical temperature of 5 K and a maximum inversion temperature of about 50 K. The boiling point of liquid helium is 4.2 K at normal pressure. The problem of producing large quantities of liquid helium was first solved by Collins, Fig. 5.3. The pure helium gas is compressed to a pressure of 17 atmospheres at a temperature of 77 K using liquid nitrogen as a coolant. The liquid nitrogen may be obtained from air liquefied using a Stirling engine refrigerator, as briefly discussed in Section 3.7. The gas is then cooled to 9 K using two expansion engines and delivered to the Joule–Kelvin valve at 17 atmospheres pressure. A fraction of

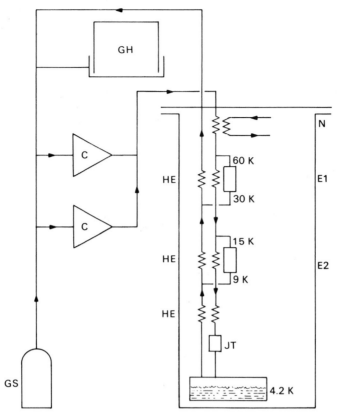

Fig. 5.3 Schematic of a helium liquefier. Pure helium gas from the gas holder (GS) is passed through the compressors (C) and emerges at a pressure of 17 atmospheres. After passing through the liquid nitrogen cooled heat exchanger (N) the gas passes to two expansion engines (E1, E2) and a Joule-Kelvin expansion valve (JT). The expanded gas passes back to the gas holder (GH) via the heat exchangers (HE). The liquid helium is held in a Dewar vessel.

the helium then liquefies at a pressure of 1.3 atmospheres. The remainder of the cold helium gas is used to cool the incoming gas in a counter-current heat exchanger. The advantage of the Joule–Kelvin expansion is that it requires no moving parts although the temperature drop for a given pressure drop is less than that for a reversible adiabatic expansion (Exercise 5.6). Modern machines do not actually require a supply of liquid nitrogen but the through put is doubled if it is available.

Low-temperature physics only became possible with the development of helium liquefiers. Many experiments are simply carried out with the sample immersed in liquid helium at 4.2 K, but by reducing the pressure above the liquid using a large vacuum pump the temperature can be reduced to about 1.1 K. At 2.17 K normal helium (^{4}He) becomes a superfluid (Section 10.1) and faster pumping does not lower the temperature much below 1.1 K because of the very large thermal conductivity of ^{4}He in this region. The rare isotope of helium, ^{3}He remains a normal fluid to much lower temperature (Sections 10.1 and 10.4) and may be used to reduce the temperature to about 0.3 K. A refrigerator using mixtures of ^{3}He and ^{4}He may also be used down to temperatures of about 0.01 K (10 mK). Methods of reaching still lower temperatures are discussed in Sections 10.4 and 10.5.

5.4 Phase transitions

The isotherms of a pure substance were shown in Fig. 2.6 as a function of pressure and volume. The equivalent isotherms as a function of pressure and density are shown in Fig. 5.4. At high temperature and low density the isotherms are identical to those of a perfect gas but below the critical temperature T_c it is possible to liquefy the gas by applying sufficiently high pressure. The flat region of the isotherms (along which liquid and vapour are in equilibrium) reduces to a point of inflexion on the critical isotherm. The critical isochore is shown by a dotted line in Fig. 5.4.

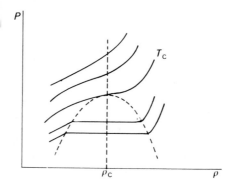

Fig. 5.4   The isotherms for gas and liquid as a function of pressure and density. The critical isochore is shown as a dotted (vertical) line.

The pressure–temperature diagram is also of interest (Fig. 5.5). Vapour and liquid are in equilibrium along the line ab but above T_c it is not possible to convert vapour to liquid by an increase of pressure and therefore the line stops at b. Along oa vapour and solid are in equilibrium. The solid–liquid equilibrium along ac does not appear to have a critical temperature so, unlike ab, ac extends without limit. The intersection of the three curves represents the triple point on the P–T diagram.

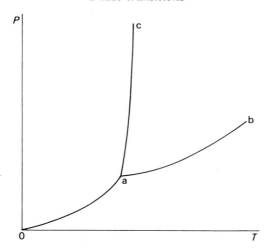

Fig. 5.5 The pressure–temperature diagram for phase equilibria.

5.4.1 *First-order phase transition: Clapeyron's equation*

The solid–liquid, solid–vapour and liquid–vapour (except at the critical point) phase transitions are accompanied by changes of density or, equivalently, volume per unit mass (v) and by a latent heat (l). The latent heat is related to the change of entropy of the system by eqn 3.26

$$s_2 - s_1 = \frac{l}{T}$$

where T is the temperature of the transition and s is the entropy per unit mass of a phase.

The first derivatives of the Gibbs free energy for unit mass therefore changes discontinuously since

$$\left(\frac{\partial g}{\partial T}\right)_P = -s \quad \text{and} \quad \left(\frac{\partial g}{\partial P}\right)_T = v \qquad (5.25)$$

and the phase transition is called first order. The classification of phase transitions by the lowest derivatives of the Gibbs function which changes discontinuously is due to Ehrenfest. The liquid–vapour transition at the critical temperature and pressure involves no change in volume and is therefore an example of a higher-order phase transition (Section 10.4.2).

The equation of the lines shown in Fig 5.5 was first derived by Clapeyron. A phase change occurs at constant pressure and temperature and under these conditions it has already been shown (Chapter 4) that the Gibbs function for the system must be a minimum. When n_1 moles of phase 1 exist in equilibrium with n_2 moles of phase 2 the total Gibbs function is

$$G = n_1 g_1 + n_2 g_2$$

where as usual g refers to unit mass. A change of n_1 by dn requires a charge of $-$dn in n_2 to conserve the total mass of the system and therefore

$$dG = (g_1 - g_2) \, dn$$

dG must be zero at equilibrium, since G is required to be a minimum, and therefore $g_1 = g_2$.

The Gibbs functions per unit mass of each phase in equilibrium are equal, i.e. the chemical potentials (Section 4.5) of each phase are equal.

A small change of temperature and pressure *along the coexistence curve* will lead to equal changes in g_1 and g_2

$$\left(\frac{\partial g_1}{\partial T}\right)_P dT + \left(\frac{\partial g_1}{\partial P}\right)_T dP = \left(\frac{\partial g_2}{\partial T}\right)_P dT + \left(\frac{\partial g_2}{\partial P}\right)_T dP$$

which using eqn 5.25 becomes

$$\left(\frac{\partial P}{\partial T}\right)_\sigma = \frac{s_2 - s_1}{v_2 - v_1} \tag{5.26}$$

or since the latent heat $l = (s_2 - s_1) T$

$$\left(\frac{\partial P}{\partial T}\right)_\sigma = \frac{l}{T(v_2 - v_1)} \tag{5.27}$$

where the subscript σ means that the differential is to be evaluated along the coexistence curve. It should be remembered that s, l and v must all refer to the same mass of substance, v is therefore the reciprocal of the density.

A normal solid is more ordered, has lower entropy, than the corresponding liquid and l is therefore positive, that is heat is needed to change a solid at its melting point to a liquid. However an exception to this rule is discussed in Section 10.4.1. The latent heat is never negative for the liquid–vapour transition.

The sign of the slope of the liquid–solid coexistence curve, ac in Fig. 5.5 is determined by the density of the liquid and solid phases at the melting point. The solid is usually denser than the liquid and therefore dP/dT is positive, as shown in Fig. 5.5. A well-known exception to this behaviour is water, for which dP/dT is negative, since ice is less dense than water, and ice therefore tends to melt under increased pressure.

5.4.2 The liquid-vapour transition. Integration of Clapeyron's equation

An approximate relationship between the pressure and boiling point of a liquid may be derived by integrating eqn 5.27. The density of a vapour, away from the critical temperature, is much less than that of its liquid ($v_v \gg v_l$). Assuming that the equation of state of the vapour may be approximated by the perfect gas law, eqn 5.27 becomes

$$\frac{\mathrm{d}P}{\mathrm{d}T} = \frac{l}{Tv_v} = \frac{lP}{R_m T^2}$$

Then if l is independent of temperature

$$\ln\left(\frac{P}{P_0}\right) = \frac{l}{R_m}\left(\frac{1}{T_0} - \frac{1}{T}\right). \tag{5.28}$$

The vapour pressure therefore varies exponentially with temperature. Equation 5.28 is useful for rough calculations but is improved by using a virial expansion for the function of state and also by allowing for the temperature dependence of the latent heat. The simple form

$$l = l_0 - aT$$

where l_0 and a are constants, is often used.

Exercises

5.1. Find the constant A in eqn 5.12 for copper at 300 K using the following data: Molar volume 7.06×10^{-6} m^3. $C_P = 24.5$ J K^{-1} mol^{-1}

$$\alpha = 4.92 \times 10^{-5} \text{ K}^{-1} \qquad \kappa_T = 7.76 \times 10^{-12} \text{ N}^{-1}\text{ m}^2$$

At 800 K $C_P = 28.0$ J K^{-1}. Find the value of C_V at 300 K and 800 K.

5.2. Find the change in temperature of 1 mole of a van der Waals gas after a free expansion from 10^{-3} m^3 to 2×10^{-3} m^3. Take $a = 0.14$ N m^4 mol^{-2}.

5.3. Show that equation 5.21 for the change of temperature in a Joule–Kelvin expansion is equivalent to

$$\Delta T = -\frac{1}{C_P}\left\{\left(\frac{\partial U}{\partial P}\right)_T + \left[\frac{\partial(PV)}{\partial P}\right]_T\right\}\Delta P$$

The temperature change of a gas in the Joule–Kelvin expansion is due to the *sum* of the deviations from the two perfect gas laws.

5.4. Derive eqn 5.22 and hence find the change of temperature of a van der Waals gas at the Boyle temperature after a Joule–Kelvin expansion.

5.5. The virial equation may be written:

$$PV = RT + B_P P + C_P P^2 + D_P P^3 + \ldots$$

Find the equation of the inversion curve and show that an equation involving only B_P and C_P is inadequate at low temperatures. (The results of Exercise 2.2 relating B_P and C_P to the van der Waals constants may be used.)

5.6. Show that for a given pressure change the cooling in a Joule–Kelvin expansion is always less than that due to a reversible adiabatic expansion.

5.7. In the Grüneisen model of a solid the entropy is a function of x only where $x = \theta/T$ and θ, a characteristic temperature for the solid, is a function of volume only. Express $(\partial S/\partial V)_T$ and $(\partial S/\partial T)_V$ in terms of dS/dx and hence show that in this model

$$\alpha V = \gamma_G \kappa_T C_V = \gamma_G \kappa_S C_P$$

$$\gamma_G = -d \ln \theta / \partial \ln V = V(\partial P/\partial U)_V$$

γ_G is called the Grüneisen constant for the solid and is between 1 and 3 for most metals.

5.8. Show that the Mie-Grüneisen equation of state,

$$(P - P_0)V = \gamma_G(U - U_0)$$

where P_0 and U_0 are the values at 0 K, follows from the definition of γ_G in the previous question.

5.9. The Anderson-Grüneisen parameter (δ_{AG}) is an important quantity in the physics of materials at high temperatures and pressure. It is defined by the equation

$$\delta_{AG}(T,P) = -\frac{1}{\alpha B_T}\left(\frac{\partial B_T}{\partial T}\right)_P$$

where the coefficient of thermal expansion and the isothermal bulk modulus (the reciprocal of the isothermal compressibility) are functions of both temperature and pressure. The value of δ_{AG} is about 3 for many oxides. (a) Show from the result of Exercise 5.7 that the temperature dependence of B_T may be deduced from experimental values of C_V if δ_{AG} and γ_G is known. (b) Derive the general thermodynamic equality

$$\left(\frac{\partial \alpha}{\partial P}\right)_T = \frac{1}{B_T^2}\left(\frac{\partial B_T}{\partial T}\right)_P$$

and hence show that

$$\left(\frac{\partial \ln \alpha}{\partial P}\right)_T = -\frac{\delta_{AG}}{B_T}$$

(c) Show that

$$\frac{\alpha(T,P)}{\alpha(T,0)} = \left[\frac{V(T,P)}{V(T,0)}\right]^{\delta_{AG}}$$

if δ_{AG} is independent of pressure
(A more general treatment is given in Guillermet, (1986)).

5.10. At low temperature the heat capacity at constant volume of an insul-ator is proportional to T^3. How does $C_P - C_V$ vary with temperature?

5.11. Sketch the isotherms on a P–V diagram for a vapour in equilibrium with its liquid. Consider a Carnot engine working between T and $T + \Delta T$ in the coexistence region. Show, that if 1 mole of liquid is converted to vapour at $T + \Delta T$, the work done in the cycle is $dP(V_V - V_1)$. Hence derive the Clausius–Clapeyron equation, eqn 5.27 (This was the original deriv-ation of the equation.)

5.12. The molar volume of ice at 273.15 K and 1 atmosphere pressure is 19.6 cm^3 and that of water 18.00 cm^3. Show that the melting point of ice decreases under pressure. The latent heat of fusion of ice is 6.0 kJ mol^{-1}. Find the pressure to reduce the melting point by 1 K. (The melting point of most liquids increases under pressure. The agreement between experiment and theory for water was an important step in establishing the validity of thermodynamics.)

5.13. Trouton's rule, a useful approximation to the behaviour of many liquids, states that the ratio of the latent heat of evaporation to the temperature of the normal boiling point is equal to 88 J K^{-1} mol^{-1}. Find the vapour pressure of zinc at 800 °C if the normal boiling point is 907 °C.

5.14. Find the change in the melting point of gallium under a pressure of 0.1 GPa (1 kbar) given that the normal melting point is 30 °C, the latent heat of fusion is 77 J g^{-1} and the density of solid (liquid) gallium is 5.89 (6.08) g cm^{-3}.

Conclusion to Part I

Thermodynamics has been seen in the last two chapters to be of importance both for pure science and for its applications. A language has been constructed which can describe the equilibrium states of any system and can also predict the observed transitions between the various states which satisfy the law of the conservation of energy. The application of the second and third laws of thermodynamics together with the Maxwell relations and the mathematics of functions of two variables has led to general results for such practical problems as the reduction of measurements to constant volume and the liquefaction of gases. Further examples of the applications of thermodynamics will be given in Chapter 10, after a microscopic view has been developed in Part II.

Thermodynamics can be extended greatly beyond the range of problems that have been considered so far in this book. The extension to systems of more than two variables which has only been briefly considered in this book is of fundamental importance in chemistry where the number of molecules of a given type in the system must be treated as a variable. Similarly the thermodynamics of *small* systems where the ratio of surface to volume properties is not negligible, or very large systems where gravitational effects cannot be neglected, are of great importance in biology and astronomy respectively.

Thermodynamics may also be extended (if certain further assumptions are introduced) to systems which are not in thermal equilibrium. The flow of heat across a finite temperature gradient for example has been seen to be an irreversible process. The theory of the transport properties (such as thermal conductivity) of a system is therefore sometimes called the thermodynamics of irreversible processes and has been an active branch of theoretical physics for many years. Transport theory is considered from a simpler viewpoint in Chapter 9 but the thermodynamic theory will not be considered in this book.

The limitation of thermodynamics that always remains however, is that the fundamental equation from which all the equilibrium properties of the system can be calculated has to be obtained from *outside* thermodynamics. The fundamental equation for a given system could be established by an enormous amount of experimental work. This approach may sometimes be necessary in an engineering problem, but is unsatisfactory in science. What is required is an atomic theory for the system of interest from which the fundamental equation can be deduced. All the equilibrium properties of the system can then be evaluated using thermodynamics alone.

The microscopic (atomic) theory of the equilibrium properties of a system (called statistical mechanics) will be developed in the next section. Statistical mechanics also provides new insights into thermal processes. For example the entropy of a system, although well defined in terms of thermodynamics, may have seemed a rather mysterious quantity. What are the consequences at the atomic level of an increase in the entropy of a system?

Thermodynamics can provide a clue to the answer to this question but the complete answer requires the atomic viewpoint of statistical mechanics. As an example of the thermodynamic argument consider the following information obtained from X-ray analysis of rubber fibres: when the rubber fibre is stretched it is found to be in the crystalline state. The atoms therefore form a regular *ordered* array. When the tension is removed the rubber forms an amorphous (disordered) material. This change may be correlated with the entropy of the rubber fibre as a function of tension by using the Maxwell relation

$$\left(\frac{\partial l}{\partial T}\right)_{\mathscr{F}} = \left(\frac{\partial S}{\partial \mathscr{F}}\right)_{T}$$

obtained by substituting l for V and $-\mathscr{F}$ for P in eqn 4.13. The left-hand side of the equation is found by experiment to be *negative* for rubber, that is to say the length of a fibre held at constant tension *decreases* as the temperature is raised. This is of course the opposite change to that normally found, for example in a steel wire. The entropy of rubber when stretched at constant temperature therefore *decreases* and the degree of atomic order (amorphous to crystalline solid) *increases*. (The statistical mechanics of rubber is discussed in Section 7.4.)

If we are prepared to generalise this observation to all systems, then the second law of thermodynamics in the form: 'the entropy of a system within adiabatic walls cannot decrease' might be written: 'the degree of atomic order of a system within adiabatic walls cannot increase'. The system cannot *spontaneously* pass to a state of higher order. (The rubber fibre was of course not within adiabatic walls but in contact with a heat reservoir at temperature T so its entropy could both increase and decrease.)

This connection between entropy and atomic order will be developed in the next section, but it is worth stressing that the thermodynamic entropy is a completely *macroscopic* quantity (defined to within a constant by $dS = dQ_R/T$) independent of any microscopic theory. An 'atomic' entropy must therefore be shown to have the same properties as the macroscopic entropy. The connection between entropy and atomic order is very difficult to establish for all but the simplest systems, but the macroscopic entropy remains a useful parameter.

PART II
Equilibrium statistical mechanics

6

Weakly coupled systems

The discussion in Section 2.3 showed that thermodynamic concepts such as temperature are not required for the discussion of the behaviour of a single particle in a conservative system, since the energy of the particle is constant. When the isolated system consists of a number of particles (such as molecules of a gas) however, the individual particle energies change after each collision in such a way that the energy of the entire system (E) is conserved.

$$E = \varepsilon_1 + \varepsilon_2 + \ldots + \varepsilon_N.$$

In general it is no longer possible (or indeed interesting) to discuss the exact behaviour of each molecule. In thermal equilibrium the macroscopic properties of the system become independent of time and at the molecular level it is postulated that the *fraction* of the molecules with a given range of energies is also independent of time. An individual molecule will leave the chosen energy range after a collision but elsewhere in the gas another molecule will on average also make a collision such that it enters the energy range.

The concept of a distribution function to describe the equilibrium properties of a classical perfect gas was first introduced by Maxwell and was later extended by Boltzmann. The distribution function $f_0(v, r)$ is defined such that: $f_0(v, r)\mathrm{d}^3v\mathrm{d}^3r$ is the number of molecules of the gas in the velocity range v to $v + \mathrm{d}v$ and with position between r and $r + \mathrm{d}r$ when the gas is in the thermal equilibrium.

Velocity is a vector quantity involving both magnitude (speed) and direction. In many books however no distinction is made in the text between speed and velocity and it is then necessary to consider whether the vector or scalar quantity is under consideration. The symbols d^3v, d^3r are used to mean a three-dimensional volume element such as $\mathrm{d}v_x\,\mathrm{d}v_y\,\mathrm{d}v_z$ in cartesian coordinates or $v^2\,\mathrm{d}v\,\sin\theta\,\mathrm{d}\theta\,\mathrm{d}\phi$ in polar coordinates as convenient.

The general distribution function can often be simplified. When gravitational effects are neglected the energy of a molecule of a perfect gas is independent of position. The integral of $f_0(v, r)$ over all velocities and over the allowed volume of the gas (which is simply equal to N, the total number of molecules) is then separable

$$\int \int f_0(v, r) \mathrm{d}^3 v \mathrm{d}^3 r = \int f_0(v) \mathrm{d}^3 v \int \mathrm{d}^3 r = N.$$

Therefore

$$\int f_0(v) \mathrm{d}^3 v = \frac{N}{V}.$$

The correct form for the velocity distribution function $f_0(v)$ was first given by Maxwell by an ingenious but inadequate argument (Exercise 6.1) and first derived rigorously by Boltzmann for the case of binary collisions of gas molecules

$$f_0(v) \mathrm{d}^3 v = \frac{N}{V} \left(\frac{m}{2\pi kT} \right)^{\frac{3}{2}} \mathrm{e}^{-\frac{mv^2}{2kT}} \mathrm{d}^3 v \qquad (6.1)$$

This equation will be derived as a special case of the general results of statistical mechanics in Section 8.1. In eqn 6.1, m is the mass of the molecule, k is the Boltzmann constant and T the absolute temperature.

An alternative interpretation of eqn 6.1 is that the *probability* of a given molecule having velocity in the range v to $v + \mathrm{d}v$ is given by

$$\left(\frac{m}{2\pi kT} \right)^{\frac{3}{2}} \mathrm{e}^{-\frac{mv^2}{2kT}} \mathrm{d}^3 v. \qquad (6.2)$$

The transition between the properties of a single molecule (which can only be described statistically by eqn 6.2) and the whole system of N particles is only possible because in a perfect gas the molecules are on average well separated and *weakly interacting*. This separation would not be possible in a liquid, where the properties of the system as a whole must always be considered.

The fundamental concept of thermodynamics was found to be the entropy function. The fundamental concept of classical statistical mechanics is the distribution function. The relationship between these two quantities was first established by Boltzmann who defined a function H such that

$$H = \int f(v, r, t) \ln f(v, r, t) \mathrm{d}^3 v \mathrm{d}^3 r. \qquad (6.3)$$

Boltzmann showed that as a result of binary collisions in the gas an arbitrary function $f(v, r, t)$ always changed in time until it came into the Maxwell form given by eqn 6.1. The function H always *decreased* with time to a minimum value which was associated with thermal equilibrium. A connection is therefore suggested with the (negative) of the entropy of the system since the entropy cannot decrease for a system within adiabatic walls. The form

$$S_B = -kVH_0 \qquad (6.4)$$

where H_0 is the function of eqn 6.4 with $f_0(v, r)$ given by eqn 6.1 may be shown directly (Exercise 6.2) to give the entropy of a perfect gas in agreement with thermodynamics.

A number of objections were advanced against the *H* theorem and it was subsequently modified by Boltzmann and others (see Tolman, 1938 for example). Consider, for example, that d^3v represents the limit of some small volume Δ^3v. The number of particles in Δ^3v may be calculated as a function of time by computer experiments (Section 1.2) and is found to *fluctuate* even for times greater than the relaxation time of the system. The value of *H* therefore also fluctuates, as shown in Fig. 1.2. The question of fluctuations is considered later (Section 7.1) but it is sufficient for the moment to state that since the importance of fluctuations in the classical region decreases as the inverse square root of the number of molecules, they will normally be unimportant for macroscopic systems containing perhaps 10^{20} particles.

The classical Boltzmann entropy function was extended to include quantised systems by Planck. The approach via quantum mechanics is in fact simpler than the earlier classical argument and will be used in this book. A perfect gas is one example of a weakly interacting system of particles. These systems are of great importance because of their relative simplicity and will be defined and discussed in the next two sections. The Boltzmann–Planck entropy will then be calculated for a number of systems and will be shown to lead to the fundamental equation (in the sense of Chapter 4) from which all the thermodynamic properties of the system may be calculated.

6.1 Systems of identical particles

The present section is intended to form a very brief revision of some of the concepts of quantum mechanics, particularly as applied to systems of identical particles. The reader who is unfamiliar with these concepts may still follow the argument of Sections 6.3 and 6.4 but would need to consult a text on quantum mechanics before reading Section 6.5.

According to quantum mechanics, the Schrödinger wave equation of a system contains all the information available about that system. This information may however be less than would be expected from the classical physics of macroscopic bodies. For example the two electrons in the helium atom are considered to be indistinguishable by experiment, whereas the moons of a planet are certainly distinguishable.

The time-independent Schrödinger equation

$$\mathcal{H}\psi = E\psi$$

where $\mathcal{H}$ is the Hamiltonian operator and *E* the energy, has in general both discrete (quantised) energy solutions and also regions in which the energy is continuous. The electronic energy levels of the hydrogen atom are shown in Fig. 6.1. The energy levels are quantised but come closer and closer together at higher energies.

An independent solution of the wave equation is called a *state* of the system, or a microstate if it is thought necessary to distinguish it from the

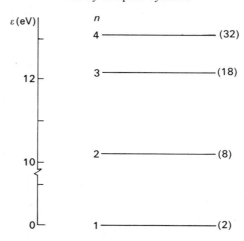

Fig. 6.1 The lower allowed electronic energy levels of the hydrogen atom. The degeneracy of each level is shown at the right. The degeneracy is equal to $2n^2$ where n is the principle quantum number.

thermodynamic state (macrostate) discussed in Part I. In general a number (g) of independent solutions of the wave equation have the same energy (E_n). The energy level n is then said to be g-fold degenerate. In many cases it will be seen that it is not necessary to consider the effect of very small splittings of the energy levels. If the separation between a number of levels is much less than the thermal energy (kT) they may be grouped and treated as one degenerate level.

The wave function of a free particle in space has the form

$$\psi = Ae^{i\mathbf{k}\cdot\mathbf{r}}$$

(where A is a normalising constant, $k = 2\pi/\lambda$ and λ is the de Broglie wavelength of the particle) and the wave equation has only a continuous energy solution. However if the particle is considered to be confined to a box of volume V the energy levels become quantised, although for a macroscopic box (say of side 0.01 m) the levels are extremely close together (Appendix IV and Exercise 6.3) relative to the thermal energy kT. The allowed energy states may then be treated as a continuous function (the density of states) such that

$$D(\varepsilon)d\varepsilon = \text{number of allowed states in the}$$
$$\text{energy range } \varepsilon \text{ to } \varepsilon + d\varepsilon$$

The density of states function for the molecules of a gas in fact provides the link between the quantum and classical theories of perfect gases.

When the system of interest consists of more than one identical particle, the wave functions of the individual particles will in general overlap and the

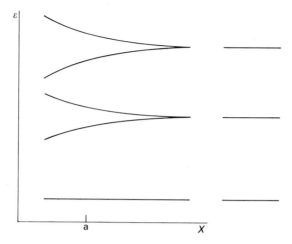

Fig. 6.2 The energy levels of the isolated atom (right) are broadened into bands when the wave functions overlap as in a crystalline solid. A typical lattice spacing (a) is shown.

particles will interact so strongly that the allowed energy levels of the whole system will become quite different from the energy levels in the individual particles. The energy levels of the electrons in a crystalline solid are shown schematically in Fig. 6.2. When the atoms are widely separated (as in a gas) each atom has the same set of electronic energy states. As the distance between the atoms decreases the sharp atomic energy levels spread out into *bands* of allowed energy separated by regions (gaps) in which no energy states exist. The width of a band depends upon the degree of overlap of the atomic wave functions. In general then, there is no simple relationship between the wave function of the whole (strongly interacting) system and that of the separated atoms.

It is however possible to say something about the behaviour of the wave function of any system if two identical particles with overlapping wave functions are interchanged. Consider for simplicity a system of just two identical particles in a nondegenerate state. Then if the coordinates of the two particles are written A and B respectively, where the coordinates are the three spacial dimensions plus the spin of the particle

$$\psi(A, B) = \pm \psi(B, A) \tag{6.5}$$

When the sign is positive the wave function is called symmetrical, when negative antisymmetrical.

The restriction imposed on the wave function by eqn 6.1 arises because it is not possible to *distinguish* between the two particles. The probability that one particle is in a volume dV_A at A and the other in a volume dV_B at B is given by

$|\psi(A, B)|^2 \, dV_A \, dV_B$. Since it is not possible to say which of the particles is at A and which at B the interchange of the two particles does not lead to a new state of the system and $|\psi(A, B)|^2$ is equal to $|\psi(B, A)|^2$.

It may be shown that a system in a state described by a wave function of one symmetry can never pass to a state described by a wave function of the opposite symmetry.

The fundamental particles are therefore divided into two groups which may be shown to be characterised by the spin of the particle.

Particles with integral or zero spin (for example photons, ^{4}He) have symmetrical wave functions. They obey Bose–Einstein statistics (Section 6.5) and are often called bosons.

Particles with half-integral spin (for example electrons, protons, neutrons, ^{3}He) have antisymmetrical wave functions. They obey Fermi–Dirac statistics (Section 6.5) and are often called fermions.

The Pauli exclusion principle arises directly from the requirement that the wave function of a system of particles with half-integral spin be antisymmetric. The wave function is found to vanish if two particles are in the same single-particle quantum state (including spin). A statement of the principle is: only one fermion can be placed in each single-particle state of the system.

If the spin states are considered separately then $(2s + 1)$ fermions can be found in each energy state. Therefore a maximum of two electrons $(s = \tfrac{1}{2})$ can be found in each energy state when the state is defined without spin.

In general the Pauli exclusion principle requires that the properties of a gas of fermions be quite different from those of the classical Maxwell gas (as might be expected) but it is also found that a boson gas may be quite different from both the classical and fermion gases. It might therefore appear that two or three perfect-gas laws are required, but we shall see that at the densities and temperatures at which gases exist the quantum corrections to the classical equation of state $(PV = RT)$ are usually negligible (Section 7.5).

6.2 Weakly coupled systems

When N identical particles are brought together to form a macroscopic system they will in general interact strongly (as discussed in the last section). A particularly simple situation arises however where the interaction of the particles is so weak that the allowed energy levels of each particle are unchanged. This situation could arise in two rather different ways.

A *localized* system of particles could be established with a sufficiently large separation between the particles for the overlap of their wave functions to be neglected. The interaction must be just sufficient for the system as a whole to come to a common temperature. In this case each particle interacts independently with an external stimulus (say a magnetic field) and the N-body problem reduces to a single-body problem. The localized, weakly interacting particles must be treated as *distinguishable* since the sites of the particles may be labelled.

The separation of the atoms in a solid is not sufficient for the whole system to be treated as a collection of independent particles but it will be seen that certain aspects of the solid state can be treated (either exactly or to a first approximation) within the weakly coupled model. A nucleus with spin $\frac{1}{2}$, for example, has a very small magnetic moment whose interaction with a neighbouring nucleus may be neglected relative to the interaction with a magnetic field. The nuclear magnetism may therefore be treated as the interaction of N weakly coupled nuclei with the magnetic field. As a further example, the lattice dynamics of a solid can be treated, to a rough approximation, as equivalent to a set of N independent simple harmonic oscillators (the Einstein model, Section 10.3).

When the wave functions of a set of *non-localized* particles are allowed to overlap the particles must be treated as indistinguishable. When the energy of interaction between the particles is small the wave function of the whole system may be considered to be formed from the products of the wave functions of the individual particles, and can only be symmetrical or antisymmetrical depending upon the spin of the particles, as discussed in Section 6.1. The particles are then distributed over the allowed single-particle energy states, allowing for the exclusion principle if the particles are fermions. The single-particle energy states must be found using quantum mechanics.

The present section may be summarized as follows. The wave function of a macroscopic system is a function of all the particles in the system. This is an N-body problem to which there is in general no exact solution but in a weakly coupled system the following information is required.

1. The allowed energy states of a *single* particle.
2. Whether or not the particles are localized. Localized particles will be treated as distinguishable, non-localized (identical) particles as indistinguishable.
3. If the particles are non-localized they must be classified as either bosons or fermions. The restriction that no two fermions can be in the same single-particle state must be observed.
4. The macroscopic state of the system (for example constant energy and volume) must be specified.

The methods of statistical mechanics will now be applied to a particularly simple localized system (in Section 6.3), and to a gas (in Section 6.5). The general results of statistical mechanics will be discussed in Chapter 7.

6.3 Two model systems

The simplest quantized system to discuss is an assembly of N identical weakly interacting *localized* particles, each of which can only exist in one of two non-

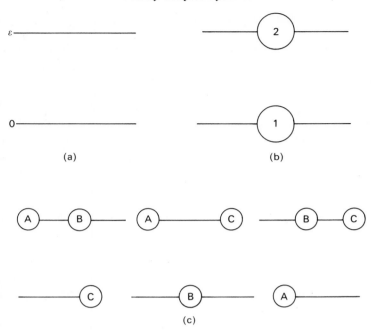

Fig. 6.3 (a) A two-level system. A particle can exist only in the state of zero energy or with energy ε. (b) The energy of the system (the macrostate $E = 2\varepsilon$) is specified by the occupation numbers. (c) The microstate of the system requires the specification of the energy level occupied by each particle. To each macrostate there corresponds a finite number of microstates.

degenerate energy levels separated by an energy ε [Figs. 6.3(a) and (b)]. The application of this model to a real system is discussed in Chapters 7 and 10.

Since the particles are localized they are to be treated as distinguishable and the interchange of two particles between the two energy levels leads to a new microstate of the system, although the energy of the system is unchanged. The energy of the whole (isolated) system (E) and the number of particles (N) in volume (V) defines the macrostate of the system. The distinguishable microstate for a given (N, E, V) can then be enumerated.

The microstates for $N = 3$ and $E = 2$ are shown in Fig. 6.3(c). Notice that the macrostate is defined by just the number of particles in each energy level (the occupation numbers) but the microstate requires the specification of the particular particles in each level because the localized particles must be treated as distinguishable. The number of microstates for a given macrostate therefore depends strongly upon the distinguishability of the particles.

In order to go beyond the representation of the microstates shown in Fig. 6.3 it is essential to make some assumption about the *probability* with which each microstate occurs. The fundamental assumption of statistical mechanics (the assumption of *a priori* probabilities) is: 'a closed system (that is with N, E, and V constant) in thermal equilibrium is equally likely to be in any of the microstates accessible to it'. A microstate is said to be accessible if it satisfies the macroscopic conditions imposed on the system (such as $N = 3$, $E = 2$ in Fig. 6.3). The microstate refers to the system as a whole. In the case of weakly coupled systems the accessible states of the whole system may be constructed using both the macroscopic conditions and any restriction on the filling of the single-particle states (Section 6.4).

The justification of the assumption of *a priori* probabilities must ultimately rest upon the agreement between the calculations of statistical mechanics and experiment. It is in fact the simplest assumption that can be made and is sometimes justified in terms of our ignorance of the detailed behaviour of atomic systems.

If each of the microstates of the model system is treated as having equal probability, an equation for the number of distinguishable microstates for a given macrostate may be found immediately. The energy of the system can only take on the value $M\varepsilon$ where M is the number of particles in the upper energy level. There are then $(N - M)$ particles in the lower level. The total number of permutations of N particles is $N! = 1 \times 2 \times 3 \times \ldots \times (N-1) \times N)$ but the *order* in which the identifiable particles go into the energy levels does not change the microstate. The total number of microstates $[W(E)]$ for energy $M\varepsilon$ is therefore

$$W(E) = \frac{N!}{M!(N-M)!} \tag{6.6}$$

$$E = M\varepsilon$$

There is nothing more to say about the microstates of the system so a connection between $W(E)$ and the entropy of the system must be sought. The first statement of this connection for a quantised system was given by Planck, following the classical statistical mechanics of Boltzmann (eqn 6.4)

$$S_B = k \ln W(E) \tag{6.7}$$

where k is the Boltzmann constant. The form S_B is used to stress that it must still be shown that this function is equivalent to the thermodynamic entropy (S). The Boltzmann–Planck entropy is correctly extensive since if two identical systems S_1 and S_2 are joined together

$$S = S_1 + S_2 \tag{6.8}$$

but the number of states accessible to the new system is

$$W = W_1 W_2 \tag{6.9}$$

The form of eqn 6.7 satisfies eqns 6.8 and 6.9. As was remarked at the beginning of this section however, the Boltzmann–Planck entropy allows the entropy of an isolated system to fluctuate and therefore decrease from its equilibrium value (although the fluctuations are normally negligible). This point will be discussed further in Chapter 7.

The fundamental equation of the model system has therefore been obtained (if it is agreed to accept the entropy defined by eqn 6.7 as equivalent to the thermodynamic entropy).

$$S_B = k(\ln N! - \ln M! - \ln (N - M)!). \tag{6.10}$$

An important simplification of this equation is obtained if it is remembered that N is normally very large ($\approx 10^{24}$ particles) and that for large numbers

$$\ln N! \approx N \ln N - N \quad \text{(Stirling's approximation)}. \tag{6.11}$$

If M and $(M - N)$ are also taken as being very large, the equation becomes

$$S = k[N \ln N - N - M \ln M + M - (N - M) \ln (N - M) + N - M]$$
$$= k[N \ln N - M \ln M - (N - M) \ln (N - M)]. \tag{6.12}$$

The temperature of a macroscopic system is defined (from the second law of thermodynamics)

$$T\,dS = dE + P\,dV$$

$$\frac{1}{T} = \left(\frac{\partial S}{\partial E}\right)_V = \frac{1}{\varepsilon}\left(\frac{\partial S}{\partial M}\right)_V \tag{6.13}$$

where the total energy (E) has been taken as identical to the thermodynamic internal energy. Applying eqns 6.12 and 6.13

$$\frac{1}{T} = \frac{k}{\varepsilon}[\ln (N - M) - \ln M]$$

or

$$M = (N - M)e^{-\frac{\varepsilon}{kT}}. \tag{6.14}$$

Therefore

$$M = \frac{N}{1 + e^{\varepsilon/kT}} \tag{6.15}$$

$$E = \frac{N\varepsilon}{1 + e^{\varepsilon/kT}} \tag{6.16}$$

The entropy of the system at temperature T may be found by substituting for M and $(N - M)$ in eqn 6.12.

Notice that as the temperature goes to absolute zero the energy and the entropy both go to zero. The model system is therefore in agreement with the third law of thermodynamics, which in terms of eqn 6.7 may be reinterpreted:

'there is only one allowed microstate of a system at absolute zero'. It is widely believed (although there is in fact no proof) that the true ground state of any system is non-degenerate. In fact so long as the ground state is only g-fold degenerate (rather than g^N-fold) the contribution to the entropy would be vanishingly small due to the fact that the entropy depends upon $\ln W(E)$ (see eqn 6.20).

There are three other features of the model calculation which should be noted because of their general importance. The original system was defined to be an isolated system of constant energy. An expression was then found for the number of microstates which satisfied the specified energy (E) and the entropy of the system was defined by eqn 6.10. The *temperature* however could only be introduced once the assumption had been made that the number of particles in the system was large. 'A large isolated system acts as its own thermal reservoir and a temperature may be defined for the system.'

The second point to notice is that the calculation using Stirling's approximation, while justifiable for large N, also required M and $(N-M)$ to be large. At low temperatures M is not large and it might therefore be thought that eqn 6.15 was only approximately correct. This is not in fact the case, as will be seen later, if M is taken to be the *thermal average* number of particles in the upper state at temperature T. That is to say the temperature is treated as the fundamental quantity, rather than E. The energy defined by eqn 6.16 is then the average energy at temperature T and fluctuations about the mean energy are now allowed (Section 7.1).

The third point is that expressions like 'high temperature' and 'low temperature' have a definite meaning for quantized systems. The important parameter is not the temperature as such but the ratio ε/kT, as may be seen from eqn 6.15 or 6.16. A characteristic temperature T_g may be defined in terms of the separation of the allowed energy levels.

$$\varepsilon = kT_g. \tag{6.17}$$

At the temperature $T \gg T_g$ and from equation 6.15 $M = N/2$. It is left to the reader to show that in this high temperature limit

$$S = Nk\ln 2 \tag{6.18}$$

The specific heat associated with a two-level system therefore goes to zero at both high temperature and low temperature (where S goes to zero) since

$$C_V = T\left(\frac{\partial S}{\partial T}\right)_V.$$

The two-level system is rather unusual in that most quantised systems do not have an energy level beyond which the energy cannot rise. One of the consequences of this maximum in the energy is discussed in Section 7.8.

A simple harmonic oscillator has a set of energy levels given by $\varepsilon = (n + \frac{1}{2})\varepsilon_0$ where n is an integer greater than or equal to zero and $\varepsilon_0 = h\nu$ where ν is the frequency of the oscillator and *h* the Planck constant. The levels therefore extend to higher energies without limit. The allowed microstate for a set of *N* weakly interacting simple harmonic oscillators with total energy *E* may be found by exactly the same method as for the two-level system although the results are rather more complicated. It is convenient to take the energy zero at $\varepsilon_0/2$. The allowed energy levels are then simply a ladder of spacing ε_0 [Fig. 6.4(a)].

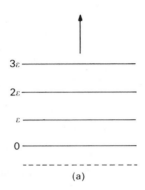

(a)

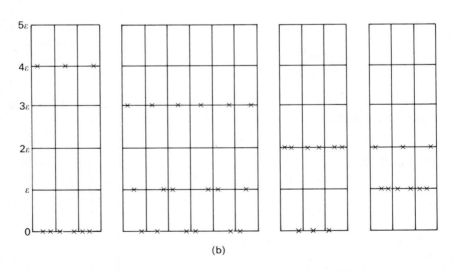

(b)

Fig. 6.4   (a) The allowed energy levels of a simple harmonic oscillator extend without limit to higher energies; The zero-point energy is shown as a broken line. (b) The allowed microstates for $N = 3$, $E = 4\varepsilon$.

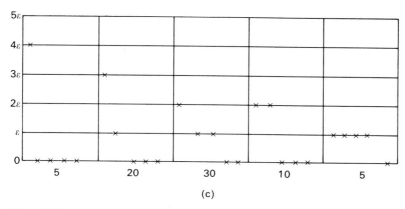

(c)

Fig. 6.4 (c) The occupation numbers for $N = 5$, $E = 4\varepsilon$ and the number of times they occur.

The allowed microstates for three particles and a total energy of $4\varepsilon_0$ are shown in Fig. 6.4(b). The occupation numbers of the energy levels are seen to occur in three or six ways. When the number of particles is increased to five with total energy $4\varepsilon_0$ a very important feature (characteristic of large systems) begins to appear [Fig. 6.4(c)]. One particular set of occupation numbers (2,2,1,0,0) occurs more frequently than any other set and the distribution of the occupation numbers over the energy levels in this most probable set is also close to that in the next most probable (3,1,0,1,0). For the simple harmonic oscillator it is still possible to write an *exact* expression for the number of microstates (Exercise 6.4) and to calculate the occupation of the levels at temperature T as was done for the two-level system. In general however this exact calculation is not possible, as will be seen in the next section and for the case of a gas in Section 6.5.

The solution to this problem is to note that as the number of particles increases the most probable distribution of occupation numbers becomes completely dominant. Furthermore since even for five particles the probability of occupation of an energy level within the most probable arrangement is close to the probability calculated over all the microstates (Exercise 6.5) it will be sufficient for a large system to calculate only the distribution within the most probable arrangement. The entropy is then defined by the approximate expression

$$S = k \ln W_{\text{MP}} \tag{6.19}$$

where W_{MP} means the number of microstates in the most probable arrangement.

The distinction between eqn 6.19 and the original expression for the entropy (eqn 6.7) is unimportant for large systems because S depends on $\ln W$.

As an example consider the high-temperature limit of the two-level system; then $S = k \ln W(E) = Nk \ln 2$. The number of microstates available to the system is (with N typically $\approx 10^{24}$ for a macroscopic system)

$$e^{N \ln 2} \approx 10^{\frac{N}{3}} \approx 10^{10^{24}}.$$

This is a number compared to which even numbers like 10^{10} are completely insignificant. Once the logarithm has been taken, the error in the entropy caused by neglecting 10^{10} states is

$$\delta S \approx k \ln 10^{10} \approx 23k \tag{6.20}$$

but the entropy is

$$S \approx 10^{24} k \tag{6.21}$$

In the discussion of the general localized weakly coupled system in the next section we will therefore set out to find the distribution of particles over the energy levels which occurs in the maximum number of ways. This most probable distribution will be considered to be effectively identical to the actual distribution for a large system.

6.4 The general weakly coupled localized system

The general weakly coupled localized system may be defined as follows: the system consists of N distinguishable particles each of which may exist in a set of energy levels of energy ε_r with degeneracy g_r. The total energy of the isolated system is E and both E and N remain constant. The number of particles in each energy level at any instant must therefore satisfy the two sum-rules

$$E = n_1 \varepsilon_1 + n_2 \varepsilon_2 + \ldots = \sum_r n_r \varepsilon_r \tag{6.22}$$

$$N = n_1 + n_2 + \ldots = \sum_r n_r. \tag{6.23}$$

The number of microstates accessible to the system if the energy levels were nondegenerate would be given by a generalization of the two-level result (eqn 6.6)

$$W(E) = \frac{N!}{n_1! n_2! \ldots n_r! \ldots} \tag{6.24}$$

When an energy level is g_r-fold degenerate however and contains n_r distinguishable particles, the particles can be distributed among the microstates in $g_r^{n_r}$ ways. The total number of microstates is therefore

$$W(E) = N! \Pi_r \frac{1}{n_r!} g_r^{n_r} \tag{6.25}$$

where it must be remembered that the n_r are restricted by the two sum-rules given in eqns 6.22 and 6.23.

The most probable values for n_r may now be found by maximizing equation 6.25 subject to the restrictions of 6.22 and 6.23. Taking the logarithm of eqn 6.25 and using Stirling's approximation (eqn 6.11)

$$\ln W(E) = \ln N! + \sum_r n_r \ln g_r - \sum_r \ln n_r!$$

$$= N \ln N - N + \sum_r n_r \ln g_r - \sum_r (n_r \ln n_r - n_r)$$

$$= N \ln N + \sum_r n_r (\ln g_r - \ln n_r).$$

The maximum value of $\ln W(E)$ will occur for the distribution of the n_r distinguishable particles for which $\delta \ln W(E)$ is zero.

$$-\delta \ln W(E) = -\sum_r (\ln g_r - 1 - \ln n_r)\delta n_r = 0. \tag{6.26}$$

The function $\ln W(E)$ is here being approximated by a continuous function. The variation in $\ln W(E)$ must take place subject to the two conditions

$$\delta N = \sum_r \delta n_r = 0 \tag{6.27}$$

$$\delta E = \sum_r \varepsilon_r \delta n_r = 0. \tag{6.28}$$

Therefore eqn 6.26 becomes

$$\sum_r (\ln n_r - \ln g_r)\delta n_r = 0. \tag{6.29}$$

This problem of a maximum subject to certain restraints may be handled by the method of Lagrange undetermined multipliers (Appendix III). The eqn 6.27 is multiplied by some factor (α), the eqn 6.28 by some factor (β) and the two equations added to eqn 6.29. Then

$$\sum_r (\ln n_r - \ln g_r + \alpha + \beta\varepsilon_r)\delta n_r = 0. \tag{6.30}$$

The effect of the undetermined multipliers is to make the δn_r independent. In this case the only solution to the equation is for the bracket in eqn 6.30 to be zero for *all* r. Therefore

$$n_r = g_r e^{-(\alpha + \beta\varepsilon_r)} \tag{6.31}$$

where α and β, the undetermined multipliers, have still to be found. Equation 6.31 provides the solution to the most probable number of particles in the energy level r of degeneracy g_r. The most probable number of particles in one single-particle state is therefore simply

$$n_r = e^{-(\alpha + \beta \varepsilon_r)}. \tag{6.32}$$

It will be shown in Section 7.3 that β is given by

$$\beta = \frac{1}{kT} \tag{6.33}$$

where k is, as usual, the Boltzmann constant and T the absolute temperature. The constant α is obviously determined at a given temperature by the condition that the total number of particles be N. Therefore

$$\sum_r n_r = N = e^{-\alpha} \sum_r e^{-\beta \varepsilon_r} \tag{6.34}$$

where the *sum runs over the single particle states* of the system or equivalently

$$N = e^{-\alpha} \sum_r g_r e^{-\beta \varepsilon_r} \tag{6.35}$$

where the *sum is over the single particle energy levels* of the system. The complete most probable distribution of particles over the allowed states of the system can therefore be found if the allowed states are known. The results for the two-level system may be verified directly (Exercise 6.7).

The method used in this section to derive the most probable distribution of particles over the allowed energy states is extended to perfect gases in the next section, but it is perhaps already obvious that its application is limited to weakly coupled systems. In Chapter 7 the generalization of statistical mechanics to all systems in thermal equilibrium will be discussed and the approximate result for the most probable number of particles in a state (eqn 6.32) will be seen to be the exact result for the thermal average number of particles in the state.

6.5 A gas of weakly coupled particles

The distribution of the velocities of a perfect gas given in eqn 6.1 was derived by Maxwell using the methods of classical physics. The form of the distribution function is correct but it really the limiting case of the quantum-mechanical distributions which will now be considered. In Section 6.1 it was observed that there are two types of particles. Those with integral or zero spin have symmetrical wave functions and will be called bosons. In a gas the particles are to be treated as indistinguishable but there is no restriction on the number of bosons in a single-particle state. The particles with half-integer spin (fermions) are also indistinguishable but their antisymmetric wave functions lead to a maximum of one particle per single-particle state (the Pauli exclusion principle). The counting of distinguishable microstates for a given macrostate is therefore different from the results for localized particles discussed in Sections 6.3 and 6.4.

The allowed energy states of a particle in a box are derived in Appendix IV. The states are extremely close together for a box of macroscopic dimensions and may be treated in terms of a continuous function (the density of states) defined by eqn A4.31. The counting of the microstates for a given macrostate is therefore performed by considering the number of energy states in some small range of energy $\Delta\varepsilon_r$ to be

$$g_r = D(\varepsilon)\,\Delta\varepsilon_r. \tag{6.36}$$

The range $\Delta\varepsilon_r$ must be large enough to make g_r large [since it will be necessary to use Stirling's approximation (eqn 6.11)] but small enough not to affect the calculation of macroscopic quantities. However since the example given by eqns 6.32 and 6.33 showed that macroscopic quantities like the entropy are rather insensitive to small changes at the microscopic level it would be expected that the restriction on $\Delta\varepsilon_r$ is not serious.

The macroscopic state of the gas is defined by the number of particles (N), the volume (V) and the allowed energy of the isolated gas (E). In fact it is only necessary to say that the energy is defined within some value δE, rather than the non-physical value of exactly E, since this small uncertainty will again (as in the case of the entropy) not affect the results for a large system.

The macroscopic conditions are therefore summarized by the equations

$$N = \sum_r n_r; \quad \delta N = \sum_r \delta n_r = 0 \tag{6.37}$$

$$E = \sum_r n_r\varepsilon_r; \quad \delta E = \sum_r \varepsilon_r\delta n_r = 0 \tag{6.38}$$

in exactly the same way as for the localized system.

The number of distinguishable microstates for n_r indistinguishable particles distributed over g_r states is however quite different from eqn 6.25 where the localized particles were treated as distinguishable, and also depends upon whether the particles are bosons or fermions.

In a gas of bosons there is no restriction on the number of particles in a single state. The n_r particles could be divided into g_r groups by placing $(g_r - 1)$ partitions between the particles (Fig. 6.5). The total number of objects under consideration is now $(n_r + g_r - 1)$. The total number of permutations is $(n_r + g_r - 1)!$ but the permutations of the n_r indistinguishable particles $(n_r!)$ and the $(g_r - 1)$ indistinguishable partitions $(g_r - 1)!$ do not lead to a new microstate. The total number of distinguishable microstates is therefore

$$\frac{(n_r + g_r - 1)!}{n_r!(g_r - 1)!} \tag{6.39}$$

for the group of states r and

$$W_{\mathrm{B}} = \Pi_r \frac{(n_r + g_r - 1)!}{n_r!(g_r - 1)!} \tag{6.40}$$

Fig. 6.5 Bose–Einstein counting. The particles can be divided into four groups by inserting three partitions. Therefore in general into g groups by inserting $(g-1)$ partitions.

for the whole system. The argument leading to eqn 6.39 is sometimes found rather difficult to follow. A direct calculation of the states for a small number of particles (Exercise 6.7) may give increased confidence in the correctness of the result.

In a gas of fermions a maximum of one particle is allowed to be in each single-particle state. The first particle can therefore be distributed over the g_r states in g_r ways but the second in only (g_r-1) ways and so on. If the particles were *distinguishable* the total number of ways of distributing n_r particles over g_r states would be

$$g_r(g_r-1)(g_r-2)\ldots(g_r-n_r+1)=\frac{g_r!}{(g_r-n_r)!}.$$

The number of states must of course be greater than or equal to the number of particles. Since the particles are indistinguishable the number of microstates for the system is

$$W_F=\Pi_r\frac{g_r!}{n_r!(g_r-n_r)!}. \tag{6.41}$$

The calculation of the most probable distribution can now be carried out using Stirling's approximation provided that n_r, g_r and (g_r-n_r) may all be taken as much greater than unity

$$\ln W_B=\sum_r\left[(n_r+g_r)\ln(n_r+g_r)-n_r\ln n_r-g_r\ln g_r\right]$$

$$\ln W_F=\sum_r\left[(n_r-g_r)\ln(g_r-n_r)-n_r\ln n_r+g_r\ln g_r\right] \tag{6.42}$$

$$-\delta\ln W_B=\sum_r\left[\ln n_r-\ln(n_r+g_r)\right]\delta n_r=0$$

$$-\delta\ln W_F=\sum_r\left[\ln n_r-\ln(g_r-n_r)\right]\delta n_r=0$$

where the maximum must be found subject to the conditions given by eqns 6.37 and 6.38. As in Section 6.4, two Lagrange undetermined multipliers are introduced to make the variations δn_r independent

$$-\delta \ln W_B = \sum_r [\ln n_r - \ln (n_r + g_r) + \alpha + \beta \varepsilon_r] \, \delta n_r = 0$$

$$-\delta \ln W_F = \sum_r [\ln n_r - \ln (g_r - n_r) + \alpha + \beta \varepsilon_r] \, \delta n_r = 0$$

and each term in the brackets must be zero.

Hence for bosons

$$n_r = \frac{g_r}{e^{\alpha + \beta \varepsilon_r} - 1} \tag{6.43}$$

and for fermions

$$n_r = \frac{g_r}{e^{\alpha + \beta \varepsilon_r} + 1}. \tag{6.44}$$

The eqn 6.43 was first derived independently by Bose and Einstein and eqn 6.44 by Fermi and Dirac. They are usually called the Bose–Einstein and Fermi–Dirac distribution functions and will be considered in more detail in Section 10.1. The Lagrange undetermined multiplier β is again equal to $(kT)^{-1}$ as in Section 6.4. The multiplier α, at a given temperature, is determined by the normalisation condition

$$N = \sum_r \frac{g_r}{e^{\alpha + \beta \varepsilon_r} \pm 1} \tag{6.45}$$

where the positive sign applies to fermions and the negative to bosons. This sum may be converted to an integral [because for a particle in a box the states of the system have been found to be very close together (see Appendix IV)] using the density of single-particle states function

$$N = \int_0^\infty \frac{D(\varepsilon) d\varepsilon}{e^{\alpha + \beta \varepsilon} \pm 1}. \tag{6.46}$$

The density of states function derived in Appendix IV then leads to

$$N = 2\pi V \left(\frac{2m}{h^2}\right)^{\frac{3}{2}} \int_0^\infty \frac{\varepsilon^{\frac{1}{2}} \, d\varepsilon}{e^{\alpha + \beta \varepsilon} \pm 1} \tag{6.47}$$

(ignoring the spin factor). This equation *defines* α at a given temperature, remembering that $\beta = (kT)^{-1}$ but cannot be solved in general in closed form.

The general eqns 6.45 and 6.47 will be discussed in Section 10.1, but it is useful to obtain the *classical limit* of eqn 6.47 at this point. The quantum-statistical result given in eqn 6.47 may be expected to reduce to the classical Maxwell distribution when the mass of the particles is large, the number density of the particles is small (therefore when the volume of the gas is large for a given number of particles) and the temperature of the system is high. Under these conditions it may be seen from eqn 6.47 that exp α must become much greater than one and the integral may be written

$$N = 2\pi V \left(\frac{2m}{h^2}\right)^{\frac{3}{2}} \int_0^\infty \frac{\varepsilon^{\frac{1}{2}}\,d\varepsilon}{e^{\alpha+\beta\varepsilon}}$$

$$= 2\pi V \left(\frac{2m}{h^2}\right)^{\frac{3}{2}} e^{-\alpha} \int_0^\infty \varepsilon^{\frac{1}{2}}\,e^{-\beta\varepsilon}\,d\varepsilon. \tag{6.48}$$

The two quantum distributions have therefore become identical in the classical limit, as was to be expected since all perfect gases have the same equation of state regardless of the spin of the particles which make up the gas. The value of α may be found by putting the integral into standard form using the substitution

$$x = \beta\varepsilon \quad dx = \beta\,d\varepsilon = d\varepsilon/kT$$

$$N = 2\pi V \left(\frac{2mkT}{h^2}\right)^{\frac{3}{2}} e^{-\alpha} \int_0^\infty x^{\frac{1}{2}}\,e^{-x}\,dx.$$

The integral is a gamma function with value $\sqrt{\pi}/2$ (Appendix II). Therefore

$$e^{-\alpha} = \frac{N}{V}\left(\frac{h^2}{2\pi mkT}\right)^{\frac{3}{2}} \tag{6.49}$$

The classical limit requires that exp $-\alpha \ll 1$. The physical significance of this may be seen by writing equation 6.49

$$\left(\frac{V}{N}\right)^{\frac{1}{3}} \gg \left(\frac{h^2}{2\pi mkT}\right)^{\frac{1}{2}}. \tag{6.50}$$

The left-hand side of the equation is the mean separation of the particles. The right-hand side may be shown to be of the order of the de Broglie wavelength of a typical particle from the equations

$$p = \frac{h}{\lambda}$$

$$\varepsilon = \tfrac{1}{2}mv^2 = \frac{p^2}{2m} = \frac{h^2}{2m\lambda^2}.$$

The mean energy of a gas molecule in a monatomic classical gas is $3kT/2$ (Section 7.5) so

$$\lambda = \frac{h}{\sqrt{(2m\varepsilon)}} \approx \frac{h}{\sqrt{(3mkT)}} \tag{6.51}$$

which is comparable with the right-hand side of eqn 6.50. Quantum effects are therefore unimportant when the separation of the molecules is much greater than their de Broglie wavelength.

The conclusion drawn from eqn 6.50 is satisfied by real gases at normal temperature and pressure (Exercise 6.8). The deviations from the perfect gas law $PV = RT$ discussed in Section 2.2 are due to interactions between the molecules rather than to quantum effects associated with $\exp -\alpha$ becoming small. (See also Section 10.1.)

The conclusion drawn from eqn 6.50 that quantum effects are unimportant when the mean separation of the particles is much greater than their de Broglie wavelength is physically reasonable on the basis of wave mechanics, since under these conditions the 'particle' rather than the 'wave' aspect of the molecule will be dominant.

An alternative way to view the transition from the two quantum gases to the single classical gas is to note that the exclusion principle will not have any practical effect on the selection of states for fermions if there are far more states available to the system than particles to fill them. The number of states per unit energy range increases with the volume of the box and the mean energy of the system increases with temperature. A combination of a low particle number-density and high temperature would therefore be expected to be favourable to the classical limit, in agreement with eqn 6.50. When there are far more states available than particles ($g_r \gg n_r$) eqns 6.40 and 6.41 *both* reduce to

$$W = \Pi_r \frac{1}{n_r!} g_r^{n_r} \tag{6.52}$$

The most probable number of particles in the single particle state r is now

$$n_r = e^{-(\alpha + \beta \varepsilon_r)} \tag{6.53}$$

in agreement with the earlier calculation for *localised* particles but the number of distinguishable microstates is a factor of $N!$ *less* than the number for localized particles (eqn 6.25) because the gas particles must be treated as indistinguishable. In the earlier classical theory due to Maxwell and Boltzmann all particles were treated as distinguishable so eqn 6.52 is sometimes called the *corrected* Maxwell–Boltzmann counting of microstates for non-localised particles.

The equilibrium properties of the quantum and classical gases could now be developed and in particular the identity $\beta = (kT)^{-1}$ established. This will be left until Section 7.3 however after the general results of statistical mechanics have been obtained.

6.6 Conclusion

The approach to statistical mechanics discussed in this chapter may have been felt to be rather unsatisfactory. The use of Stirling's approximation for

example when not *all* the terms in the equation are very large is mathematically unjustifiable and must involve the physical assumption that the
neglect of small terms has no effect on the calculation of the macroscopic
properties of the system. This was shown to be plausible for the entropy in the
high-temperature limit where S is of order Nk but is more difficult to justify at
low temperature where the entropy goes to some small value or to zero.

The second curious feature of the calculations in this chapter is that the
original macrostate was specified in terms of constant energy (to within some
small range of energy) but once the system had been taken to be macroscopic
it was possible to use the Boltzmann–Planck equation $S = k \ln W(E)$ and the
thermodynamic equation

$$\frac{1}{T} = \left(\frac{\partial S}{\partial U}\right)_V = \left(\frac{\partial S}{\partial \overline{E}}\right)_V$$

to define a *temperature* for the model system. Similarly in the more general
cases of weakly coupled particles discussed in Sections 6.4 and 6.5 a Lagrange
undetermined multiplier (β) had to be introduced which can be shown to be
equal to $1/kT$ by first deriving the perfect gas law $P = nkT$ and then recog-
nizing that in mutual thermal equilibrium all systems must have the same
value of T (Section 2.1). This transition required the identity of the Boltzmann
entropy with the thermodynamic entropy, and of the total energy of the
system with the thermodynamic internal energy. The approximation involved here (from using Stirling's equation to find the entropy) becomes
smaller as the size of the system increases.

The limit in which

$$N \to \infty, \ V \to \infty, \ \left(\frac{N}{V}\right)_{\to n} \tag{6.54}$$

is called the thermodynamic limit. The thermodynamic theory presented in
Part I also involved the above limit since, for example, the internal energy
was treated as an extensive quantity ($U = uV$) which is only justifiable if the
atoms near the surface of the sample form a negligible fraction of the whole.
Thermodynamics and statistical mechanics should therefore lead to exactly
the same results in the limit given by eqn 6.54.

The introduction of the temperature of a system as a derived property of a
large isolated system is rather in conflict with the view of temperature as a
fundamental thermodynamic quantity. A real experimental system is never
completely isolated from its environment, more usually it is in contact with a
heat reservoir at temperature T. In Chapter 4 the transition from a thermodynamic isolated system to one at constant volume and temperature was
found to lead to the Helmholtz free energy (rather than the entropy) as the
fundamental thermodynamic quantity. Similarly in the next chapter the
relationship $S = k \ln W(E)$ for an isolated system will be shown to be replaced

by an equation which is most simply expressed in terms of the Helmholtz free energy.

The most unsatisfactory aspect of the method presented in this chapter however is that it is quite impossible to perform calculations for systems which are not weakly coupled. A strongly interacting system must be treated as a whole. The total energy E cannot be broken up into individual particle energies. As an example consider the interaction of the (N) molecules of a dense gas. The energy might be written

$$E = \tfrac{1}{2}m \sum_{r=1}^{N} v_r^2 + U(r_1, r_2 \ldots r_N).$$

The kinetic energy of the molecules is separable into single-particle energies but the potential energy of a given molecule depends upon the position of all the other molecules. Only when the potential energy term is zero (as in a perfect gas) is the total energy of the system separable into a sum of single-particle energies.

The extension of the Boltzmann–Planck entropy concept to all systems, first made by Gibbs, is discussed in the next chapter and used to derive the general relationship between the atomic states of a system and thermodynamics. The general equations however can only be evaluated approximately for most strongly interacting systems.

Exercises

6.1. Maxwell suggested that a function $f(v_x)dv_x$ existed for a perfect gas in thermal equilibrium such that $Nf(v_x)dv_x$ was the number of molecules with velocity between v_x and $v_x + dv_x$. He then assumed (a) the total distribution function for all directions can only be a function of the *magnitude* of the velocity (b) the velocity v_x of a molecule is independent of the value of v_y and v_z. Show that the distribution function is then of the form $A \exp Bv_x^2$. What is the weakness of this derivation?

6.2. Use the Maxwell distribution of velocities (eqn 6.1) and the Boltzmann equation $S = -kVH_0$ to derive the entropy of a perfect gas.

6.3. Calculate the energy of the first excited state of an electron in a box of side 0.01 m. A thermal energy may be defined by $\varepsilon_1 = kT_1$. Find the equivalent temperature for the first excited state. Repeat the calculation for a hydrogen molecule.

6.4. Find a general expression for the number of distinguishable microstates of N weakly coupled simple harmonic oscillators with individual energy $\varepsilon = n\varepsilon_0$ when the total energy is E.

6.5. Find the probability $P(\varepsilon)$ of a particle being on each energy level in Fig. 6.4(c). Compare this probability with the probability in the most probable arrangement.

6.6. A crystal contains N atoms, n of which, at temperature T, have been displaced from the body of the crystal to sites on the surface. Show that in thermal equilibrium, $n \approx N \exp(-\beta\varepsilon_0)$, where ε_0 is the energy required to remove one atom, if $(1 \ll n \ll N)$.

6.7. Draw all the distinguishable microstates for two particles distributed over three states for (a) distinguishable particles (b) bosons (c) fermions and show the total is in agreement with the general equations given in the text.

6.8. Show that $\exp -\alpha \ll 1$ in eqn 6.49 for helium at NTP. In silver one electron per atom may be treated as free to move through the crystal. Consider these electrons as a gas and show that $\exp -\alpha$ is not less than 1.

6.9. Show that the expression for the energy of a 2 level system, eqn 6.16, becomes

$$E = N\varepsilon \bigg/ \left(\frac{g_1}{g_2} e^{\varepsilon/kT} + 1 \right)$$

when the degeneracy of the ground state is g_1 and that of the excited state g_2.

6.10 What is the fractional error in the value of $\ln n!$ given by using eqn 6.11 when n is (a) 10 or (b) 20?

6.11. Show that $S = N k \ln 2$ for a two level system in the high temperature limit and sketch the curve of entropy as a function of temperature.

6.12. Show that the quantum mechanical results for bosons and fermions, eqns 6.40 and 6.41, reduce to the corrected classical limit eqn 6.52 when $g_r \gg n_r$.

6.13. Show that the mean de Broglie wavelength of a helium atom at NTP is comparable to its diameter but much less than the mean separation of the atoms.

6.14. The expressions 'high temperature limit' and 'low temperature limit' were defined in Section 6.3. It is important to understand that different contributions to the energy of an atom or molecule may be in different limits at a given temperature. Consider a hydrogen atom at room temperature in a box of side 0.01 m. Show that the translational (kinetic) energy is in the high temperature limit but the electronic energy is in the low temperature limit.

6.15. N molecules of a gas at temperature T are contained in a cylinder of height h and cross-sectional area A. The cylinder is in the earth's gravitational field. Show that for one molecule: (a) the probability of being found at a height z is proportional to $e^{-mgz/kT}$; (b) the mean gravitational energy is kT for a long cylinder and $\frac{1}{2}mgh$ for a short cylinder, and; (c) the r.m.s fluctuation in the gravitational energy is kT for a long cylinder. (d) Define 'short' and 'long' cylinder for this problem. (e) Show also that the pressure is an exponential function of z and find expressions for the pressure on the base of the cylinder in the 'short' and 'long' limits.

7

Equilibrium statistical mechanics

In this chapter we wish to consider a general system, with a fixed number of particles, in thermal equilibrium. (The modification to the theory when the number of particles is variable is discussed in the appendix at the end of the chapter, Section 7.10.) The system may either be isolated from its environment (and therefore at constant energy) or be in contact with a thermal reservoir at temperature T. The system at constant energy will in fact never be completely isolated and should therefore be specified as having energy in the range E to $E + \delta E$. The energy of the whole system of interest will be written as E. When weakly coupled systems are under consideration, the single-particle energy will be written ε.

The system will not remain in one particular quantum state indefinitely even when in thermal equilibrium but rather will pass through all the quantum states accessible to it as the result of small fluctuations in the surroundings. It is never possible for example to shield a system from gravitational effects. The weakly coupled system discussed in Chapter 6 would in time pass through all the states shown in Fig. 6.3 and 6.4.

Since macroscopic properties of a system in thermal equilibrium are independent of time it shoud in principle be possible to determine the probability of the system being in a particular quantum state (r) by making a series of measurements at successive times, provided that the time between measurements was sufficiently long relative to the longest relaxation time of the system. If the state r was found to occur m_r times in M measurements the probability of the state r would be defined as

$$P_r = \frac{m_r}{M}; \ M \to \infty \tag{7.1}$$

$$\sum_r P_r = \frac{1}{M} \sum_r m_r = 1. \tag{7.2}$$

The limit in eqn 7.1 is necessary since the true probabilities cannot depend upon the number of measurements made. Equation 7.2 simply means that the total probability of the system being in some quantum state is unity. Once the values of P_r were known, say at temperature T, for a given system, the macroscopic properties could be deduced as averages over P_r. The mean energy at temperature T would be

$$\bar{E} = \sum_r P_r E_r \tag{7.3}$$

where E_r is the energy of the system in the state r. The time-average value of the energy given by eqn 7.3 could now be identified (for a large system) with the thermodynamic internal energy of the system.

7.1. Ensemble averages

The method of taking time-averages over the properties of a single system does not lend itself to calculation in general, although a computer model may be used for small numbers of particles. An alternative approach to the problem was first advanced by Gibbs in 1901, using the methods of classical physics. A simplified version of the argument for quantum systems will be given here.

The single experimental system with specified macroscopic conditions is replaced for purposes of calculation by M replicas of itself. The whole collection is called an *ensemble*. The M members of the ensemble are considered to be distributed randomly over the allowed quantum states of the system. The time-average of eqn 7.1 is now replaced by an average over all the members of the ensemble at a single time. The averages taken in Chapter 6 were therefore essentially ensemble averages at constant energy for the special case of a weakly coupled system with M equal to the number of distinguishable microstates of the system.

There is no actual proof that the time-average given by eqn 7.1 and the ensemble averages taken in Chapter 6 and further discussed in this chapter will always agree. The relationship of the two averages continues to interest mathematicians but the agreement between experiment and the calculations of statistical mechanics using ensemble averages seems to be complete.

When the experimental system of interest is considered to be essentially isolated from its environment it is replaced for purposes of calculation by an ensemble of identical systems each of which is completely isolated. The energy of each system must lie within the range E to $E + \delta E$ and each system has the same volume and contains the same number of particles. Gibbs called this a *microcanonical ensemble*.

The probability of occupation of a state r of the whole system is now found using the assumption of *a priori* probabilities (Section 6.3): 'an isolated system in thermal equilibrium is equally likely to be in any of the accessible states of the system'.

The probability of the particular state r is therefore

$$P_r = \frac{1}{W(E_r)} \quad (E < E_r < E + \delta E) \tag{7.4}$$

$$= 0 \quad \text{(otherwise)} \tag{7.5}$$

where $W(E)$ is the number of states of the system with energy E or the density of states function times the energy range $D(E)\,dE$. Notice that it is the density of states of the whole system which is involved here. Only for weakly coupled systtems is it permissible to write this in terms of the density of states of individual particles.

The difficulty that arises when calculations are performed with the microcanonical ensemble is that of finding the states of the system which lie in the allowed energy range. This is more difficult than enumerating all the states of the system regardless of energy. It is for this reason that the application of the microcanonical ensemble is limited to weakly coupled systems as discussed in Chapter 6.

Two approaches are possible when the experimental system of interest contains a constant number of particles at constant volume in contact with a thermal reservoir at temperature T. The entire system of reservoir and system of interest could be treated as isolated and the properties of the composite system calculated using the microcanonical ensemble. Alternatively we may consider a set of M replicas of the experimental system placed side by side in thermal contact and imagine the whole ensemble to be isolated from its surroundings. Gibbs called this a *canonical ensemble*.

There is no restriction on the *energy* of the experimental system in this case since it may exchange energy with the thermal reservoir. Similarly in the canonical ensemble the energy of a single member of the ensemble may fluctuate by exchanging energy with its neighbours. The members of the ensemble are macroscopic and therefore distinguishable (Fig. 7.1). They are also weakly interacting because the ratio of the number of atoms near the

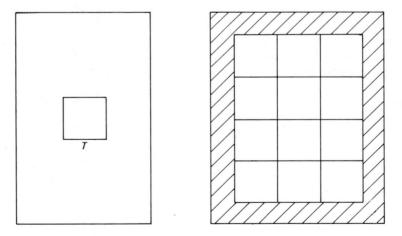

Fig. 7.1 The experimental system at temperature T is replaced for purposes of calculation by an ensemble of replicas of the experimental system which are free to exchange energy. The energy of the whole ensemble is constant.

surface to the total number in the sample is always small for a macroscopic system. The allowed energy states for each member of the ensemble are therefore identical.

The problem of M weakly coupled distinguishable objects was treated in Section 6.4. The approximate result in that case was that the probability of a particle being in the state r was given by

$$P_r = e^{-(\alpha + \beta \varepsilon_r)} \tag{7.6}$$

where $\exp - \alpha$ is given by the normalization condition

$$\sum_r P_r = 1 = e^{-\alpha} \sum_r e^{-\beta \varepsilon_r}. \tag{7.7}$$

In the present ensemble average, M may be considered to be sufficiently large for eqn 7.6 to hold exactly where the energy is now the energy of a member of the ensemble

$$P_r = e^{-(\alpha + \beta E_r)} \tag{7.8}$$

$$= \frac{e^{-\beta E_r}}{Z} \tag{7.9}$$

$$Z = \sum_r e^{-\beta E_r}. \tag{7.10}$$

The function Z is called the *partition function* or sum over states. The factor β will later be shown to be equal to $1/kT$, as in Chapter 6.

The difference between the apparently similar eqns 7.6 and 7.8 must be carefully considered. The eqn 7.8 is the exact result for the probability of occupation of the state r of the whole system at temperature T. In general there is no way of simplifying this equation. When the particles which make up the system are weakly coupled it is possible to express the partition function of the whole system in terms of the singe-particle energy states but only in the case of localized particles and the classical perfect gas does eqn 7.6 then result. The case of weakly interacting non-localized particles is discussed in Section 7.2.

The eqns 7.8–7.10 define the canonical ensemble probability of the state r of the whole system being found in an experiment. The mean energy of the system is defined as in eqn 7.3

$$\bar{E} = \sum_r P_r E_r = \frac{\sum_r E_r e^{-\beta E_r}}{\sum_r e^{-\beta E_r}} \tag{7.11}$$

where E_r is the energy of the system in state r.

In the later sections of this chapter it will become apparent that the partition function contains all the information required to calculate the thermodynamic properties of any system with a fixed number of particles, but before discussing this in detail it is of some interest to consider two points concerning ensembles is rather more detail.

The first point concerns the number (M) of members of an ensemble. In the classical theory this was simply taken to be sufficiently large not to affect the calculations, just as in the number of samples in the time average of eqn 7.1. In the quantum theory however it is possible to give a more physical meaning to the ensemble by defining it as proposed by Kittel: 'every quantum state accessible to the actual system is represented in the ensemble by one system in a stationary quantum state.' The number of states is so large, if the actual system is always thought of as either macroscopic or as a small system in contact with a thermal reservoir, that no approximation is associated with this definition. This was however essentially the method used in Chapter 6 and is therefore only approximately correct for a small system at constant energy.

The second important point is that for large systems the calculated thermodynamic properties of a system are usually independent of the type (microcanonical or canonical) of ensemble average. It was seen in Chapter 6 that a large system with energy in the range E to $E + \delta E$ behaved as its own thermal reservoir. A temperature could be defined for the system using the thermodynamic equality $1/T = (\partial S / \partial U)_V$. It will now be shown that conversely a large system at temperature T has a well-defined energy. The mean energy of the system is given by eqn 7.11. The mean square fluctuation of the energy is found using the general statistical result

$$\overline{(x - \bar{x})^2} = \overline{(x^2 + (\bar{x})^2 - 2x\,\bar{x})} = \overline{x^2} - (\bar{x})^2. \tag{7.12}$$

Therefore

$$\overline{(E - \bar{E})^2} = \frac{\sum_r E_r^2\, e^{-\beta E_r}}{Z} - \left(\frac{\sum_r E_r\, e^{-\beta E_r}}{Z} \right)^2. \tag{7.13}$$

This expression can be simplified by noting that the definition of the mean energy given in eqn 7.11 could have been written

$$\bar{E} = -\frac{1}{Z}\left(\frac{\partial Z}{\partial \beta}\right)_V = -\left(\frac{\partial \ln Z}{\partial \beta}\right)_V \tag{7.14}$$

because the states of the system are independent of the temperature. Similarly

$$\overline{E^2} = \frac{1}{Z}\left(\frac{\partial^2 Z}{\partial \beta^2}\right)_V. \tag{7.15}$$

The derivative is taken with respect to β rather than to temperature because of the simplicity of the resulting equation. The temperature derivative can of course be obtained from

$$\beta = \frac{1}{kT}, \frac{d\beta}{dT} = -\frac{1}{kT^2} = -k\beta^2. \tag{7.16}$$

The derivative of the mean energy with respect to β at constant volume is

$$\left(\frac{\partial \bar{E}}{\partial \beta}\right)_V = -\frac{1}{Z}\left(\frac{\partial^2 Z}{\partial \beta^2}\right)_V + \frac{1}{Z^2}\left(\frac{\partial Z}{\partial \beta}\right)_V^2$$

which is equivalent, using eqns 7.14 and 7.15, to eqn 7.13 apart from a change of sign, so

$$\overline{(E - \bar{E})^2} = -\left(\frac{\partial \bar{E}}{\partial \beta}\right)_V. \tag{7.17}$$

The right-hand side of eqn 7.17 may be related to the heat capacity of the system at constant volume since

$$C_V = \left(\frac{\partial U}{\partial T}\right)_V = \left(\frac{\partial \bar{E}}{\partial T}\right)_V = \left(\frac{\partial \bar{E}}{\partial \beta}\right)_V \frac{d\beta}{dT}.$$

Therefore

$$\overline{(E - \bar{E})^2} = kT^2 C_V. \tag{7.18}$$

The importance of energy fluctuations depends upon the ratio of the square root of eqn 7.18 to the mean energy of the system

$$\frac{\sqrt{\overline{(E - \bar{E})^2}}}{\bar{E}} = \frac{\sqrt{(kT^2 C_V)}}{\bar{E}}. \tag{7.19}$$

At high temperatures (the region of classical physics) the right-hand side of eqn 7.19 is always of order $1/\sqrt{N}$. The relative fluctuations about the mean energy are therefore very small for N of the order of 10^{20}. The classical gas has a mean energy of $3NkT/2$ and therefore a heat capacity at constant volume of $3Nk/2$. A solid at high temperatures has energy $3NkT$ and heat capacity $3Nk$ (Section 7.6). Fluctuations about the mean energy of a large system are only of importance at very low temperatures or when the heat capacity becomes very large, as at a phase transition (Sections 5.4 and 10.4).

The equilibrium properties of large systems with a fixed number of particles can therefore usually be calculated using either the microcanonical or the canonical ensemble. The canonical ensemble is the most convenient however because the partition function involves *all* the states of the system (weighted by the factor $e^{-\beta E_r}$) whereas the microcanonical ensemble average requires the selection from all the states of the system of just those states whose energy lies between E and $E + \delta E$.

The study of the fluctuations of a system about equilibrium does require a distinction to be made between isolated systems (whose energy cannot fluctuate more than δE) and systems with only a defined mean energy for which there is a finite (though usually very small) probability of the system passing through any energy states far from the mean energy.

A further extension of the ensemble concept may be made to systems with a variable number of particles. For example a small part of the volume of a gas could be treated as the experimental system. The rest of the gas would now act as both a heat reservoir and a particle reservoir. The experimental system is replaced for purposes of calculation by an ensemble of systems which are in thermal contact and free to exchange particles. Gibbs called this a *grand canonical ensemble*. The consequences of allowing the members of the ensemble to exchange particles should be clear following the discussion of the mean energy in the canonical ensemble. The number of particles in the system will now be variable but for a macroscopic system the fluctuations about the mean number may be shown to be small. The grand canonical ensemble may therefore be used to calculate the properties of a large closed system when the difference between N and $\bar{N}$ is unimportant, and is sometimes easier to use than the canonical ensemble.

The probability of the system being in a state r with energy E_{N_r} and a number of particles N_r is given by

$$P_r = \frac{e^{-(\alpha N_r + \beta E_{N_r})}}{\mathscr{Z}} \tag{7.20}$$

$$\mathscr{Z} = \sum_N \sum_r e^{-(\alpha N_r + \beta E_{N_r})}. \tag{7.21}$$

The function $\mathscr{Z}$ is called the grand partition function. A fuller account of the properties of the grand canonical ensemble is given in Section 7.10.

There are computational advantages in performing statistical-mechanical calculations using the grand canonical ensemble rather than the canonical ensemble for systems with a constant number of particles and the grand canonical ensemble is essential when the number of particles is allowed to fluctuate. The rest of this book however is restricted to the use of the microcanonical and canonical ensembles which are perhaps easier to visualize physically than the grand canonical ensemble. An excellent account of the use of the grand canonical ensemble is given by Kittel and Kroemer (1980).

7.2 The canonical partition function

The canonical partition function was introduced in eqns 7.9 and 7.10 as a mere normalizing factor for the probability (P_r) of a member of the canonical ensemble being in a state of energy E_r. It was apparent however from the calculation of the mean energy and the mean square fluctuation about the mean energy that thermodynamic quantities can be expressed in terms of the partition function. In the next section it will be shown that the fundamental equation for any system in thermal equilibrium with temperature T, volume

V, and containing a fixed number of particles, may be expressed in terms of the partition function. In this section the general properties of the partition function and the special case of weakly coupled systems will be considered.

The partition function expressed as

$$Z = \sum_r e^{-\beta E_r} \tag{7.22}$$

involves the sum over the distinguishable *microstates* of the whole system.

The partition function may also be expressed as a sum over the energy levels of the systems and the degeneracies of the levels

$$Z = \sum_r g_r e^{-\beta E_r}. \tag{7.23}$$

The form 7.22 will normally be used for simplicity but when actual calculations on a given system are to be performed it is necessary to remember that the degeneracies of the allowed energy levels are required (as in eqn 7.23).

When the energy levels are closely spaced relative to the thermal energy of the system it is possible to transform the sum in eqn 7.23 into an integral. The degeneracy of a given energy level is now replaced by the density of states function and the energy becomes a continuous function

$$Z = \int_0^\infty D(E) e^{-\beta E} \, dE. \tag{7.24}$$

The energy of the whole system will be a function of six variables (three for momentum and three for position) per particle, so eqn 7.24 conceals a $6N$-fold integral with N typically 10^{22}.

A more general form of eqn 7.24 may be written

$$Z = \frac{1}{h^{3N}} \int e^{-\beta E} d^{3N}p \, d^{3N}r \tag{7.25}$$

where N is the number of particles, p the momentum and r the position (Appendix IV). The notation $d^{3N}p$ is used to stress the dimension of the integral.

The physical interpretation of eqn 7.25 (as discussed in Appendix IV) is that in the $6N$-dimensional space of momentum and position (the phase space) each quantum state occupies a volume of h^{3N}. This is an example of the Heisenberg uncertainty principle which restricts the simultaneous specification of momentum and position of one particle in one dimension ($\Delta p_x \Delta x$) to order h.

The importance of the weakly interacting systems discussed in Chapter 6 should now be obvious. Only for such systems can the immense complexity hidden in the apparently simple eqns 7.22–7.25 sometimes be reduced to the relative simplicity of individual-particle states.

The partition function for N weakly coupled localized (therefore distinguishable) particles may be written in terms of the sum over single-particle states rather than the states of the whole system. A single particle has a set of allowed states with energy ε_r. The effect of the other particles is contained in the temperature of the system or equivalently in β. The partition function for a single particle (z) is

$$z = \sum_r e^{-\beta \varepsilon_r} \qquad (7.26)$$

and since the partition function for each of the N particles is identical

$$Z = z^N \quad (localized) \qquad (7.27)$$

is the relationship between the single-particle partition function and the partition function for the whole system for *localized* weakly interacting systems.

This argument *cannot* be extended to weakly interacting systems such as gases in which the wave functions of the particles overlap. The particles must now be treated as indistinguishable and it is not possible to factorise the partition functions as in eqn 7.27 except in the special case of the classical limit of the perfect gas. The effect of the particles being indistinguishable was shown to reduce the number of distinguishable states by $N!$ (eqn 6.37) in the classical limit so the relationship between the individual partition functions and that of the whole system is, from eqn 7.27.

$$Z = \frac{z^N}{N!} \quad \text{(non-localized, classical limit).} \qquad (7.28)$$

The general case of the quantum gas is best treated using the grand partition function and will only be considered in the appendix at the end of the chapter, Section 7.10.

The microstates of a system in thermal equilibrium are independent of the temperature but may be a function of the volume. The probability of a state being occupied is a function of the temperature (from eqn 7.9). The partition function is therefore in general a function of temperature and volume. The fundamental equation for any system in thermal equilibrium at given temperature and volume must be expressed in terms of the Helmholtz free energy (Section 4.1). The relationship between the Helmholtz free energy and the partition function will be discussed in the next section.

7.3 The connection with thermodynamics

In Chapter 6 the connection between the statistical concept of the number of distinguishable microstates $W(E)$ of a system with energy in the range E to $E + \delta E$ and thermodynamics was taken to be the Boltzmann–Planck equation

$$S = k \ln W(E). \qquad (7.29)$$

Applying this equation to the M replicas of the system, with M_r in the state r, which form the ensemble

$$W(E) = M! \Pi_r \frac{1}{M_r!}$$

as in Section 6.4, since the members of the ensemble are macroscopic and therefore distinguishable and using Stirling's approximation

$$S = -Mk \sum_r \left(\frac{M_r}{M}\right) \ln \left(\frac{M_r}{M}\right)$$

$$= -Mk \sum_r P_r \ln P_r.$$

The entropy of an individual member of the ensemble is therefore

$$S = -k \sum_r P_r \ln P_r \tag{7.30}$$

where the values of P_r are the appropriate ensemble probabilities for given macroscopic conditions.

In the case of an isolated system in thermal equilibrium the microcanonical ensemble leads to

$$P_r = \frac{1}{W(E)}$$

$$S = -k \sum_r P_r \ln P_r = k \ln W(E) \tag{7.31}$$

as before, since there are $W(E)$ identical terms of value $1/W(E) \times \ln W(E)$.

Equation 7.30 is sometimes taken to be the basic definition of the statistical entropy (rather than eqn 7.29) and is extended to include systems which are not in thermal equilibrium. The thermodynamic entropy is of course still restricted to systems in thermal equilibrium.

It is instructive to derive the canonical ensemble probability directly from eqn 7.30. The probability of occupation of the states r must always satisfy the normalization condition and in addition the mean energy of a system in the canonical ensemble is constant

$$\sum_r P_r = 1; \quad \sum_r P_r E_r = \bar{E} = \text{constant}. \tag{7.32}$$

The entropy of the system is defined by eqn 7.30. In thermal equilibrium the entropy of the system must be a maximum. The probability of occupation of a state r in the canonical ensemble must therefore be such that eqn 7.30 is a maximum subject to the restraints of eqn 7.32. As usual the method of Lagrange undetermined multipliers then leads to

$$\delta S = -k \sum_r (1 + \ln P_r) \delta P_r = 0$$

$$\sum_r \delta P_r = 0; \quad \sum_r E_r \delta P_r = 0$$

$$\sum_r (\alpha + \beta E_r + \ln P_r) \delta P_r = 0$$

and since each term must separately be zero

$$P_r = e^{-(\alpha + \beta E_r)}. \tag{7.33}$$

The canonical distribution therefore leads to the maximum entropy for a system with given mean energy, that is to say corresponds to the state of thermodynamic equilibrium.

The entropy defined by eqn 7.30 may be written in terms of the canonical partition function by directly substituting into the equation

$$P_r = \frac{e^{-\beta E_r}}{Z}.$$

Then

$$S = \frac{k\beta \sum_r E_r e^{-\beta E_r}}{Z} + \frac{k \ln Z \sum_r e^{-\beta E_r}}{Z} \tag{7.34}$$

$$= k\beta \bar{E} + k \ln Z$$

using the definition of the mean energy (eqn 7.11) and the normalization condition for the P_r.

It is convenient to establish the identity $kT = 1/\beta$, which has been assumed until now, at this point. From eqn (7.34)

$$\left(\frac{\partial S}{\partial \bar{E}}\right)_V = k\beta + k\bar{E}\left(\frac{\partial \beta}{\partial \bar{E}}\right)_V + k\left(\frac{\partial \ln Z}{\partial \beta}\right)_V \left(\frac{\partial \beta}{\partial \bar{E}}\right)_V$$

The left-hand side is the definition of $1/T$, where T is the thermodynamic temperature, and the last two terms on the right-hand side cancel, eqn 7.14, so

$$\beta = \frac{1}{kT}.$$

Hence,

$$S = \frac{\bar{E}}{T} + k \ln Z \tag{7.35}$$

or

$$-kT \ln Z = -TS + \bar{E}. \tag{7.36}$$

The right-hand side of the equation is simply the definition of the Helmholtz free energy if the mean energy ($\bar{E}$) is taken to be equivalent to the thermodynamic internal energy. Therefore

$$F = -kT \ln Z \tag{7.37}$$

is the fundamental equation from which all other thermodynamic functions can be derived using the results of Part I. The form of eqn 7.37 is correct since the partition function is a function of volume and temperature.

The problem of finding the fundamental equation for any system discussed in Chapter 4 has therefore become: 'evaluate the partition function of the system'. It is in fact only possible to do this exactly for a limited number of systems but other systems (such as dense gases) may be treated by methods of successive approximation similar to the perturbation theory of quantum mechanics.

The form of eqn 7.37 shows that ln Z is an extensive quantity (proportional to the number of particles in the system) as may be seen directly for the localized weakly interacting systems discussed in Section 7.2 (eqn 7.27)

$$F = -kT \ln Z = -NkT \ln z \quad \text{(localized)}. \tag{7.38}$$

The thermodynamic functions of a system in thermal equilibrium may therefore be expressed in terms of ln Z. The functions divide into two groups, (a) those involving only the derivatives of ln Z and (b) those involving ln Z. In the first group are the energy functions such as the mean energy

$$\bar{E} = \frac{\sum_r E_r e^{-\beta E_r}}{Z} = -\left(\frac{\partial \ln Z}{\partial \beta}\right)_V \tag{7.39}$$

and the specific heat at constant volume

$$C_V = \left(\frac{\partial \bar{E}}{\partial T}\right)_V = \left(\frac{\partial \bar{E}}{\partial \beta}\right)_V \frac{\partial \beta}{\mathrm{d} T} = k\beta^2 \left(\frac{\partial^2 \ln Z}{\partial \beta^2}\right)_V. \tag{7.40}$$

These quantities are unchanged if a constant term is added to the logarithm of the partition function. The distinction between the partition functions of localized and non-localized particles (eqns 7.27 and 7.28) is therefore not important when the mean energy or the specific heat are under consideration. However the group of functions which involve ln Z itself, such as the entropy and the Helmholtz free energy, do change if ln Z changes by a constant. (The distinction between eqns 7.27 and 7.28 is then of vital importance, as will be seen in Section 7.5). Suppose that the position of the energy zero is changed by an amount E_0. Then the new partition function becomes

$$Z' = \sum_r e^{-\beta(E_r + E_0)} = e^{-\beta E_0} \sum_r e^{-\beta E_r}$$
$$= Z e^{-\beta E_0}.$$

Therefore
$$\ln Z' = \ln Z - \beta E_0.$$

The mean energy becomes (from eqn 7.39)
$$\overline{E'} = \overline{E} + E_0$$

but the entropy (using eqn 7.35) is unchanged since
$$S' = k \ln Z' + k\beta \overline{E'}$$
$$= k \ln Z + k\beta \overline{E} = S.$$

The concepts introduced in this section will now be applied to the particularly simple systems of localized weakly interacting systems discussed in Chapter 6 and to the classical perfect gas. The advantages of working with the canonical ensemble rather than the microcanonical ensemble should be quickly apparent.

7.4 Localized systems

The partition function for any weakly coupled localized system involves simply the single-particle states through eqn 7.27
$$Z = z^N = \left(\sum_r e^{-\beta\varepsilon_r} \right)^N.$$

The N-particle two-level system with single-particle non-degenerate energy levels 0, ε reduces to
$$Z = z^N = (1 + e^{-\beta\varepsilon})^N$$
$$\ln Z = N \ln z = N \ln(1 + e^{-\beta\varepsilon}).$$

The number of particles in each level is then
$$N_\varepsilon = NP_\varepsilon = \frac{N\,e^{-\beta\varepsilon}}{1 + e^{-\beta\varepsilon}} \tag{7.41}$$

$$N_0 = NP_0 = \frac{N}{1 + e^{-\beta\varepsilon}} \tag{7.42}$$

$$N_\varepsilon = N_0\,e^{-\beta\varepsilon} \tag{7.43}$$

in agreement with the *approximate* calculation of the most probable number of particles in each state given in eqn 6.14. The results in eqns 7.41–7.43 are the *exact* results for the mean number of particles in each state at temperature T.

The mean energy of the system may be found from eqn 7.11
$$\overline{E} = \frac{N\varepsilon e^{-\beta\varepsilon}}{1 + e^{-\beta\varepsilon}} = \frac{N\varepsilon}{1 + e^{\beta\varepsilon}} \tag{7.44}$$

and the heat capacity at constant volume by differentiating eqn 7.44 with respect to temperature

$$C_V = Nk \left(\frac{\varepsilon}{kT} \right)^2 \frac{e^{\beta\varepsilon}}{(1 + e^{\beta\varepsilon})^2} \tag{7.45}$$

The mean energy and the heat capacity of the two-level system are shown as functions of the reduced temperature (kT/ε) in Fig. 7.2. The entropy of the two-level system may be found from either eqn 7.35 or the thermodynamic relation

$$S = -\left(\frac{\partial F}{\partial T} \right)_V = -\left(\frac{\partial F}{\partial \beta} \right)_V \left(\frac{d\beta}{dT} \right)$$

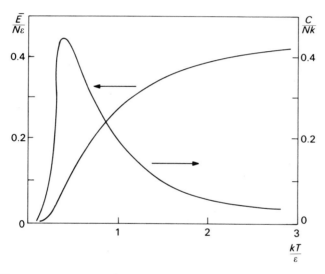

Fig. 7.2 The mean energy and heat capacity of a two-level system as a function of temperature.

where

$$F = -kT \ln Z = -\frac{N}{\beta} \ln (1 + e^{-\beta\varepsilon}).$$

Therefore

$$S = Nk \left[\ln (1 + e^{-\beta\varepsilon}) + \left(\frac{\varepsilon}{kT} \right) \frac{1}{1 + e^{\beta\varepsilon}} \right]. \tag{7.46}$$

In the low-temperature limit $(T \ll \varepsilon/k)$ the entropy approaches zero as was noted in Chapter 6 and in the high temperature limit $(T \gg \varepsilon/k)$ the entropy

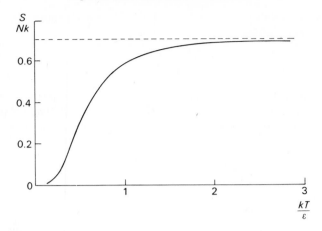

Fig. 7.3   The entropy of a two-level system as a function of temperature. The value $Nk \ln 2$ is shown as a broken line.

approaches $Nk \ln 2$. The entropy is shown as a function of reduced temperature in Fig. 7.3.

The important point, already stressed in Chapter 6, is that the properties of the system depend upon the ratio (ε/kT). The 'high-temperature limit' does not therefore mean necessarily a large number on the thermodynamic scale but simply that the energy gap (ε) is smaller than the thermal energy (kT). A quantized system with sufficiently small ε may therefore be in the high-temperature region at *all* temperatures accessible to experiment. In this case the entropy of the two-level system is $Nk \ln 2$ independent of temperature and the specific heat at constant volume, $T(\partial S/\partial T)_V$ is zero (as shown in Fig. 7.2).

The simplest example of a two-level system is a nucleus of spin $\frac{1}{2}$ in a magnetic field (Fig. 7.4). The energy level is two-fold degenerate in zero field but separates into two non-degenerate levels in a magnetic field $(\mu_0 H^*)$ with energies $\pm \mu \mu_0 H^*$ where μ is the magnetic moment of the nucleus. The energy zero is taken at zero field value and the splitting of the levels (Δ) is given by

$$\Delta = 2\mu\mu_0 H^*.$$

The single-particle partition function is now

$$z = e^{-\beta\Delta/2} + e^{\beta\Delta/2}. \tag{7.47}$$

The mean energy of the system is

$$\bar{E} = \frac{N\Delta(1 - e^{\beta\Delta})}{2(1 + e^{\beta\Delta})} \tag{7.48}$$

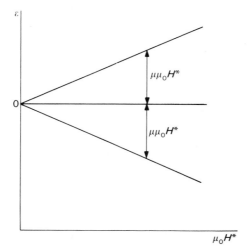

Fig. 7.4 A particle with spin 1/2 has a doubly degenerate energy level in zero magnetic field. The level splitting is proportional to the magnetic field.

and the heat capacity is unchanged from eqn 7.45. The nuclear magnetic moment is so small that in most cases the maximum magnetic field available will leave Δ/kT small at normal experimental temperature (Exercise 7.1). The specific heat is then found to be proportional to the inverse square of the temperature (Exercise 7.2). The measurement of nuclear specific heats [when they are large enough to be distinguished from other contributions to the specific heat (Section 10.3)] has been of particular importance in investigating the *internal* magnetic fields found in ferromagnetic materials such as cobalt. In these materials a large effective magnetic field is found at the nucleus even when the external field is zero.

The partition function for N weakly coupled simple harmonic oscillators is also straightforward. The energy levels of the individual oscillators are given by quantum mechanics to be

$$\varepsilon = (n + \tfrac{1}{2})\varepsilon_0; \quad \varepsilon_0 = h\nu$$

where n is an integer $(0, 1, 2, \ldots)$, h is the Planck constant and ν the frequency of the oscillator. Then the partition function for a single particle is

$$z = \sum_n e^{-(n+\frac{1}{2})\beta\varepsilon_0} = e^{-\beta\varepsilon_0/2} \sum_n e^{-n\beta\varepsilon_0}$$

$$= e^{-\beta\varepsilon_0/2}(1 + e^{-\beta\varepsilon_0} + e^{-2\beta\varepsilon_0} + \ldots).$$

The term in the bracket is an infinite geometric series with the value $1/(1 - e^{-\beta\varepsilon_0})$ so

$$z = \frac{e^{-\beta\varepsilon_0/2}}{1 - e^{-\beta\varepsilon_0}} \tag{7.49}$$

and

$$\ln Z = N \ln z = -N \left[\frac{\beta \varepsilon_0}{2} + \ln \ (1 - e^{-\beta \varepsilon_0}) \right]. \tag{7.50}$$

The mean energy is

$$\bar{E} = -\left(\frac{\partial \ln Z}{\partial \beta} \right)_V = N \varepsilon_0 \left(\tfrac{1}{2} + \frac{1}{e^{\beta \varepsilon_0} - 1} \right). \tag{7.51}$$

which is equivalent to

$$\bar{E} = \frac{N \varepsilon_0}{2} \coth \left(\frac{\beta \varepsilon_0}{2} \right). \tag{7.52}$$

At low temperature the mean energy therefore goes to $N\varepsilon_0/2$. This term arises from the condition that $\varepsilon = (n + \tfrac{1}{2}) \varepsilon_0$ rather than the value $n\varepsilon_0$ given by the old quantum theory. The term $N\varepsilon_0/2$ is called the 'zero point energy' and its effect is to preserve at low temperature the uncertainty in the position of the oscillator required by the Heisenberg uncertainty principle. In the high-temperature limit ($\beta \varepsilon_0 \ll 1$) the mean value of the energy becomes NkT. This is an example of the equipartition theorem which applies in the classical high-temperature limit (Section 7.6).

The heat capacity of the system follows immediately from eqn 7.51

$$C_V = \left(\frac{\partial E}{\partial \beta} \right)_V \frac{\partial \beta}{\mathrm{d}T} = N k \varepsilon_0^2 \beta^2 \left[\frac{e^{\beta \varepsilon_0}}{(e^{\beta \varepsilon_0} - 1)^2} \right] \tag{7.53}$$

and is shown in Fig. 7.5. The low-temperature limit ($\beta \varepsilon_0 \gg 1$) is

$$C_V = N k \varepsilon_0^2 \beta^2 e^{-\beta \varepsilon_0} \tag{7.54}$$

which decreases exponentially to zero as $T_{\to 0 \text{ K}}$. In the high-temperature limit ($\beta \varepsilon_0 \ll 1$)

$$C_V = Nk. \tag{7.55}$$

The high temperature specific heat therefore goes to a constant value (Nk) rather than to zero, as does the two-level system. This is because the two-level system cannot, at positive temperatures, have more than $N/2$ particles in the upper state (Section 7.8). Since the mean energy of the two-level system cannot increase beyond $N\varepsilon/2$, the heat capacity which involves the derivative of E goes to zero at high temperature.

The simple harmonic oscillator is of particular importance in the theory of black body radiation (Section 10.2). It also serves as a crude model for the lattice specific heat of solids (the Einstein model) in which each atom is considered to vibrate independently (Section 10.3).

One further simple example of a localized system will be considered in this section. The thermodynamic properties of rubber were seen, page 99, to be most unusual since the length of a rubber fibre at constant tension is found

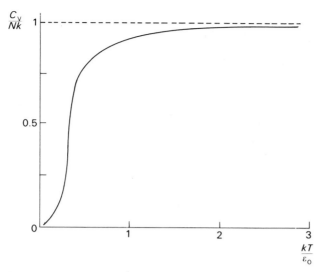

Fig. 7.5 The heat capacity of a system of weakly interacting one-dimensional simple harmonic oscillators as a function of temperature.

to *decrease* with an increase of temperature. A very simple model to account for the elastic properties of rubber considers the molecular groups as forming the links of a chain. A one-dimensional model would therefore separate the links into two groups (Fig. 7.6) which will be denoted A if they act to increase the length of the chain and B if they act to reduce it. The energy of a link is *independent* of its orientation and will be taken to be zero. This model is

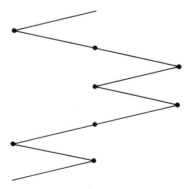

Fig. 7.6 A one-dimensional model for rubber. The links may point either to the right or left without change of energy. In the diagram the links have been displaced from the original line in order to clarify the arrangement.

plausible since rubber can be stretched to about ten times its length without exceeding the elastic limit. The partition function

$$Z = \sum_r e^{-\beta E_r}$$

involves only one term since E_r is equal to zero. The degeneracy of this term if there are N links with n_A in group A and n_B in group B is then simply the partition function

$$Z = \frac{N!}{n_A! n_B!}$$

$$N = n_A + n_B. \tag{7.56}$$

The length (L) of the chain is given by

$$L = (n_A - n_B) l$$

where l is the length of one link. The Helmholtz free energy is

$$F = -kT \ln Z = -kT (N \ln N - n_A \ln n_A - n_B \ln n_B) \tag{7.57}$$

where Stirling's approximation has been used.

The tension ($\mathscr{F}$) in the chain is given (assuming that the volume change of the fibre is negligible) by

$$\mathscr{F} = \left(\frac{\partial F}{\partial L}\right)_T = \left(\frac{\partial F}{\partial n_A}\right)_T \frac{dn_A}{dL}$$

$$= \frac{kT}{2l} \ln \left(\frac{n_A}{n_B}\right) \tag{7.58}$$

(remembering that $dn_B = -dn_A$ from eqn 7.56)

$$\mathscr{F} = \frac{kT}{2l} \ln \left(\frac{L_M + L}{L_M - L}\right)$$

where $L_M = N l$ is the maximum length of the chain. When the tension is small ($L \ll L_M$) the equation becomes

$$\mathscr{F} = \frac{kTL}{lL_M}$$

or

$$L = \text{constant} \frac{\mathscr{F}}{T}. \tag{7.59}$$

The length of the fibre therefore decreases when the temperature is raised at constant tension.

The entropy of the fibre is given (from eqn 7.57) by

$$S = k(N \ln N - n_A \ln n_A - n_B \ln n_B) \tag{7.60}$$

(remembering that $F = U - TS$ and in this case U is zero). The entropy is zero when the chain has maximum extension ($n_A = N$, $n_B = 0$) and a maximum when the extension is zero ($n_A = n_B$). The tension in the fibre arises not from any change in the internal energy with length (as in the case of a metal wire) since

$$\left(\frac{\partial U}{\partial L}\right)_T = 0$$

but from the need to overcome the entropy-driven tendency of the fibre to curl up.

This one-dimensional model of rubber cannot be expected to give exact agreement with experiment but does show the main features of the behaviour of a rubber fibre or a polymer. A three-dimensional treatment is also possible but also fails in detail because the volume change of the fibre cannot be neglected.

7.5 Classical perfect gas

The partition function for systems in which the energy levels are sufficiently close together was shown in Section 7.2 to be

$$Z = \int_0^\infty D(E) e^{-\beta E} dE$$

In the special case of the classical monatomic perfect gas the partition function may be factorized as discussed in Section 7.2 to

$$Z = \frac{z^N}{N!} \tag{7.61}$$

$$z = \int_0^\infty D(\varepsilon) e^{-\beta \varepsilon} d\varepsilon$$

where z is the single-particle partition function. The density of states function (apart from the spin factor) given in Appendix IV then leads to

$$z = \left(\frac{2m}{h^2}\right)^{\frac{3}{2}} 2\pi V \int_0^\infty \varepsilon^{\frac{1}{2}} e^{-\beta \varepsilon} d\varepsilon$$

$$= \left(\frac{2m}{h^2 \beta}\right)^{\frac{3}{2}} 2\pi V \int_0^\infty x^{\frac{1}{2}} e^{-x} dx$$

$$= V \left(\frac{2\pi m}{h^2 \beta}\right)^{\frac{3}{2}} \tag{7.62}$$

where m is the mass of one molecule of the gas. The Helmholtz free energy of the whole gas is then

$$F = -kT \ln Z = -kT \ln \left(\frac{z^N}{N!} \right)$$

$$= -NkT \left[\ln V - \tfrac{3}{2} \ln \beta + \tfrac{3}{2} \ln \left(\frac{2\pi m}{h^2} \right) - \ln N + 1 \right] \qquad (7.63)$$

using Stirling's approximation for $\ln N!$.

All the properties of the monatomic perfect gas can now be derived from eqn 7.63. The equation of state is given by

$$P = -\left(\frac{\partial F}{\partial V} \right)_T = \frac{N}{\beta V} = nkT. \qquad (7.64)$$

The mean energy may be found using eqn 7.39 or more simply using the identity (eqn 4.6)

$$U = \bar{E} = -T^2 \left[\frac{\partial (F/T)}{\partial T} \right]_V.$$

Therefore

$$\bar{E} = \tfrac{3}{2} NkT. \qquad (7.65)$$

The internal energy of a classical perfect gas is independent of the volume of the system, in agreement with Joule's (experimental) result

$$\left(\frac{\partial U}{\partial V} \right)_T = 0 \quad \text{(perfect gas)} \qquad (7.66)$$

which was used throughout Part I.

The mean energy of the perfect gas is related to the equation of state, using eqn 7.64 and 7.65, by the eqn

$$PV = \tfrac{2}{3} \bar{E} \qquad (7.67)$$

This equation is useful because it is also applicable to the Fermi–Dirac and Bose–Einstein gases (Section 10.1). The factor of 2/3 is not however correct for particles whose velocity is of the order of the speed of light, where the Einstein relativistic equation relating energy and momentum is required (Exercise 7.4).

The heat capacity of a monatomic perfect gas at constant volume is simply

$$C_V = \left(\frac{\partial \bar{E}}{\partial T} \right)_V = \tfrac{3}{2} Nk. \qquad (7.68)$$

The mean square fluctuation in the energy of a monatomic perfect gas is (from eqn 7.17)

$$\overline{(E - \bar{E})^2} = kT^2 C_V = \tfrac{3}{2} Nk^2 T^2$$

and the importance of fluctuation is , from eqn 7.19

$$\frac{[\overline{(E-\bar{E})^2}]^{\frac{1}{2}}}{\bar{E}} = \sqrt{\frac{2}{3N}}. \tag{7.69}$$

The mean energy of a classical perfect gas is therefore usually extremely well defined since N is typically $\approx 10^{20}$.

The entropy of a monatomic perfect gas (from eqn 7.63) is

$$S = -\left(\frac{\partial F}{\partial T}\right)_V = Nk\left[\frac{3}{2}\ln T + \ln\left(\frac{V}{N}\right)\right] + S_0 \tag{7.70}$$

where S_0 is a constant for a given gas

$$S_0 = Nk\left[\frac{3}{2}\ln\left(\frac{2\pi mk}{h^2}\right) + \frac{5}{2}\right]. \tag{7.71}$$

This is the Sackur–Tetrode equation for the entropy of a monatomic perfect gas in the classical limit.

The entropy of the perfect gas is in agreement with the thermodynamic result (eqn 3.41) but also contains a definite value for the constant S_0. Since S_0 contains the Planck constant it could not have been arrived at by the methods of classical physics. It is in this sense that we refer to the perfect gas as the classical limit of the quantum gases rather than as simply the classical perfect gas.

The original classical theory of the perfect gas contained an arbitrary constant instead of the Planck constant and, more seriously, used the partition function for weakly coupled distinguishable particles

$$Z = z^N.$$

The entropy given by eqn 7.70 then contains $\ln V$ rather than $\ln(V/N)$ and, as discussed in Chapter 3, this form for the entropy is not extensive and wrongly suggests that the mixing of identical gases leads to an increase in the net entropy of the system. This problem was first appreciated by Gibbs and is sometimes called the Gibbs paradox. Gibbs introduced the factor $N!$ into eqn 7.61 to bring statistical mechanics into agreement with thermodynamics, but its physical significance did not become apparent until quantum mechanics showed that the particles of the gas must be treated as indistinguishable.

The Sackur–Tetrode equation for the entropy of a perfect gas may be verified directly by experiment. The total change in the entropy of an element from the solid state near 0 K to the liquid phase and then to the vapour may be measured directly (as discussed in Chapter 3).

$$\Delta S = \int_0^{T_m} C_s \, d(\ln T) + \frac{l_f}{T_m} + \int_{T_m}^{T_b} C_1 \, d(\ln T) + \frac{l_v}{T_b} \tag{7.72}$$

where C_S is the heat capacity of the solid, T_m the melting point, l_f the latent heat of fusion, C_1 the heat capacity of the liquid, T_b the boiling point and l_v the latent heat of vaporization. The value given by experiment using equation 7.72 may then be compared directly with the calculated value from equations 7.70 and 7.71 which contain no adjustable parameters. The agreement is found to be extremely good and is perhaps the single most convincing verification of the consistency of thermodynamics and the method of ensemble averaging used is statistical mechanics.

The entropy given by eqn 7.70 is correct for a monatomic gas of particles with no nuclear or electronic spin. When the nucleus of the atom has spin I there will be $(2I + 1)$ degenerate states per atom or $(2I + 1)^N$ states for the gas. The entropy will then increase by $N \ln(2I + 1)$ but the mean energy of the system is unchanged. Similarly when the electron spin of the atom is s, the entropy will increase by $N \ln(2s + 1)$ and a gas of atoms with both nuclear spin I and atomic spin s will have a total entropy of $N[\ln(2I + 1) + \ln(2s + 1)]$ greater than that given by eqn 7.70.

The energy of a monatomic perfect gas is due to the kinetic energy of the atoms

$$E = \frac{1}{2}m \sum_{r=1}^{N} v_r^2 = \frac{1}{2m} \sum_{r=1}^{N} p_r^2$$

where p_r is the momentum of the rth atom. The only other possible contribution to the partition function is that of the electronic states of the atom but the separation between the ground state and first excited states of the atom is so large (Exercise 6.3) that at normal temperature the atom will always be in the ground state.

The energy of diatomic or polyatomic gases is due to the kinetic energy of the whole molecule and also to the internal degrees of freedom of the molecule such as rotation and vibration. The frequencies of vibration and rotation of a molecule are so different that they have little effect on each other. The partition function for the whole molecule can then be written (ignoring the spin terms)

$$z = z_t z_{rot} z_v$$

where z_t is the translational (kinetic) partition function (eqn 7.62), z_{rot} the rotational partition function and z_v the vibrational partition function. This separation follows immediately (if the energies are independent) from

$$z = \sum_r e^{-\beta \varepsilon_r} = \sum_r e^{-\beta(\varepsilon_t + \varepsilon_{rot} + \varepsilon_v)}.$$

The relationship between the single-molecule partition function and that for the whole gas of N molecules is then

$$Z = \left(\frac{z_t^N}{N!}\right) z_{rot}^N z_v^N \tag{7.73}$$

where the first term is identical with the result for a monatomic gas. The electronic partition function could also be included in the product on the right-hand side of the equation if the temperature of the system was sufficiently high for a significant number of molecules to be raised above the ground state.

The rotational energy states of a rigid system of two molecules (which is a good approximation to the behaviour of real molecules) are given by quantum mechanics to be

$$\varepsilon_{\text{rot}} = \frac{h^2}{8\pi^2 I} J(J+1) \tag{7.74}$$

where J is an integer $(0, 1, 2, \dots)$ and I the moment of inertia of the molecule about the centre of mass. The *degeneracy* of a rotational energy level is $(2J+1)$. The partition function is therefore

$$z_{\text{rot}} = \sum_{J=0}^{\infty} (2J+1) e^{-\frac{J(J+1)h^2 \beta}{8\pi^2 I}}. \tag{7.75}$$

The mean rotational energy is

$$\bar{E}_{\text{rot}} = \frac{\frac{Nh^2}{8\pi^2 I} \sum_{J=0}^{\infty} J(J+1)(2J+1) e^{-\frac{J(J+1)h^2 \beta}{8\pi^2 I}}}{\sum_{J=0}^{\infty} (2J+1) e^{-\frac{J(J+1)h^2 \beta}{8\pi^2 I}}}. \tag{7.76}$$

This expression and that resulting from the heat capacity are difficult to handle, but in the low-temperature limit $(\beta \gg 8\pi^2 I/h^2)$ it is clear that only the first term will contribute to the sum

$$\bar{E}_{\text{rot}} \approx \frac{3Nh^2}{4\pi^2 I} e^{-\frac{h^2 \beta}{4\pi^2 I}}. \tag{7.77}$$

The mean rotational energy therefore approaches zero at low temperature (where the low-temperature region is defined by $T \ll h^2/8\pi^2 kI$) and the rotational contribution to the heat capacity also goes to zero at low temperature. A gas of diatomic molecules will in principle have the same heat capacity as a monatomic gas at sufficiently low temperature but in practice this effect is only observed for hydrogen.

The importance of a characteristic temperature (related to the spacing of the energy levels by the equation $kT_g = \Delta E$) for quantized systems was discussed briefly in Section 6.3 in connection with the two-level system. It should now be clear that in general a system will have a number of characteristic temperatures such that at a given experimental temperature some aspects of the system may be in the high-temperature limit while others are in the low-temperature limit. The spacing of the translational energy states in a box of macroscopic dimensions is so small that for a gas in the classical limit these states may always be treated in the high-temperature limit and the partition function treated as an integral rather than a sum. The vibrational (and the electronic) energy states are so widely separated that at normal temperature only the ground states are occupied. These two aspects of the molecule are therefore usually in the low-temperature limit. Finally the rotational states of heavy molecules can usually be treated in the high-temperature limit, which will now be evaluated. The lightest molecules (H_2, HD) are in the transition region between the two limits below about 100 K.

The high-temperature limit of the mean rotational energy may be found by evaluating the sum in eqn 7.76 for a large number of terms but the correct result for the heat capacity may be found more simply by noting that the density of rotational states per unit energy range is given by

$$\frac{\text{Number of states}}{\text{Energy change}} = \frac{4(J+1)}{2(J+1)(h^2/8\pi^2 I)} = \text{constant}$$

(considering the degeneracies and energy separation of two adjacent states).

Therefore in the high-temperature limit ($\beta \ll 8\pi^2 I/h^2$) where the thermal energy is much greater than the separation between the energy levels, the partition function (eqn 7.75) may be transformed into an integral treating the density of rotational energy states as constant.

$$z_{\text{rot}} = \text{constant} \int_0^\infty e^{-\beta\varepsilon}\, d\varepsilon.$$

The mean rotational energy of a molecule in the high-temperature limit is then

$$\bar{\varepsilon}_{\text{rot}} = \frac{\int_0^\infty \varepsilon e^{-\beta\varepsilon}\, d\varepsilon}{\int_0^\infty e^{-\beta\varepsilon}\, d\varepsilon} = kT. \tag{7.78}$$

The rotational specific heat per molecule is then simply $(\partial\bar{\varepsilon}/\partial T)$ which is equal to k. The total heat capacity of a diatomic gas in the high-temperature limit is therefore

$$C_V = \tfrac{3}{2}Nk + Nk = \tfrac{5}{2}Nk \tag{7.79}$$

or $5R/2$ for 1 mole of the gas, but the heat capacity decreases to $3R/2$ in the low-temperature limit.

The vibrational and electronic contributions to the heat capacity are not of importance at normal temperature and will not be considered further. The rotational partition function (eqn 7.75) becomes more complicated for the case of molecules of identical atoms, and in particular for molecular hydrogen, but will not be considered further. It should be clear from this section however that the rotational and vibrational specific heats could not be understood in terms of classical physics.

The heat capacities of three very different systems have now been seen to take on the same form in the high-temperature limit. The simple harmonic oscillator $(3k)$, the free particle $(3k/2)$ and the diatomic molecule $(5k/2)$ all have heat capacities which are independent of temperature and of the same order of magnitude. These are examples of a general theorem of classical physics called the *law of equipartition of energy* which will be discussed in the next section.

7.6 The equipartition of energy

The simplest form of the equipartition theorem may be stated: 'In the classical limit a system in thermal equilibrium at temperature T has a mean energy of $kT/2$ per degree of freedom. One degree of freedom corresponds to each squared term in the expression for the energy of the system'.

A few simple examples will be discussed before proving the general theorem. The energy of one molecule of a monatomic perfect gas may be written

$$\varepsilon = \tfrac{1}{2}mv^2 = \tfrac{1}{2}m(v_x^2 + v_y^2 + v_z^2)$$

since it is free to move in three dimensions. The mean energy is therefore $3kT/2$ and the specific heat $3k/2$, in agreement with eqn 7.68.

The diatomic molecule has to be considered to have only two degrees of rotational freedom (which cannot of course be proved in classical physics) and then has the required value of $5k/2$. It is interesting to note that Maxwell and Boltzmann were not impressed by the apparent agreement between the measured specific heats of diatomic molecules and the equipartition theorem and stressed that to consider a molecule to have only two internal degrees of freedom was an unwarranted assumption. The quantum-mechanical calculation discussed in the last section should have made it clear that the other degrees of freedom are not excited at normal temperature and do not contribute to the specific heat.

The equation of a one-dimensional simple harmonic oscillator is

$$\varepsilon = \tfrac{1}{2}mv_x^2 + \tfrac{1}{2}cx^2$$

where c is a constant. In the classical limit the mean energy will be $(kT/2 + kT/2)$ since both terms contain a square. This result agrees with the high-temperature limit calculated in Section 7.4 (eqn 7.55) but at low temperature the mean energy was seen to go exponentially to the zero-point energy, not linearly to zero as suggested by the equipartition theorem.

The equipartition theorem can be stated in a more general form than that given at the beginning of this section and since the general proof is no more difficult it will be derived first and then various special cases considered. The classical partition function for the system will be written so as to distinguish a term in the energy (E') which may be a function of either momentum (p_x) or position (x). When E' is a function of momentum

$$\int e^{-\beta(E+E')} d^{3N}p \, d^{3N}r = \int' e^{-\beta E} d^{3N-1}p \, d^{3N}r \int e^{-\beta E'} dp_x \qquad (7.80)$$

where the first integral on the right-hand side excludes dp_x. The second term on the right-hand side is now integrated by parts to give

$$\int e^{-\beta E'} dp_x = \left[p_x e^{-\beta E'} \right]_{p_x=-\infty}^{p_x=\infty} + \beta \int p_x \left(\frac{dE'}{dp_x} \right) e^{-\beta E'} dp_x.$$

The first term on the right-hand side will be zero at both limits, assuming that the energy goes to infinity as the momentum goes to plus or minus infinity. Therefore

$$\frac{1}{\beta} = \frac{\int' e^{-\beta E} d^{3N-1}p \, d^{3N}r \int p_x \left(\frac{dE'}{dp_x} \right) e^{-\beta E'} dp_x}{\int e^{-\beta(E+E')} d^{3N}p \, d^{3N}r} \qquad (7.81)$$

but the right-hand side is simply a mean value so

$$\overline{p_x \left(\frac{dE'}{dp_x} \right)} = \frac{1}{\beta} = kT \qquad (7.82)$$

which may be written

$$\overline{p_x v_x} = kT \qquad (7.83)$$

since

$$\frac{dE'}{dp_x} = v_x.$$

The general result given in eqn 7.82 may be simplified when the energy is proportional to the square of the momentum

$$E' = \alpha p_x^2$$

to give simply

$$\bar{E}' = \tfrac{1}{2} kT \qquad (7.84)$$

in agreement with the first statement of the equipartition theorem.

The energy E' may refer either to a single-particle energy in a weakly coupled system (as in the examples given at the beginning of this section) or else to the energy of the system as a whole. For example the energy of a galvanometer movement with moment of inertia I suspended from a fibre with torque constant c may be written

$$E = \tfrac{1}{2} c \theta^2 + \tfrac{1}{2} I (\dot{\theta})^2 \qquad (7.85)$$

where θ is the angle of rotation and $\dot{\theta}$ the angular velocity ($d\theta/dt$). The mean energy of the suspension is therefore ($kT/2 + kT/2$) or kT.

When the energy term E' is considered as a function of position, an equation similar to eqn 7.81 may be obtained provided the energy dependence is such that on integrating by parts to obtain

$$\int e^{-\beta E'} dx = \left[xe^{-\beta E'} \right]_{x=a}^{x=b} + \beta \int x \left(\frac{dE'}{dx} \right) e^{-\beta E'} dx$$

the first term on the right-hand side vanishes at the limiting values of x. This means that the energy must increase without limit as b goes to plus infinity and a to minus infinity or that the lower limit is restricted to x equal to zero. A bound particle (such as a simple harmonic oscillator) satisfies these conditions.

The equivalent equation to eqn 7.82 when the energy E' is a function of position is then

$$\overline{x \left(\frac{dE'}{dx} \right)} = kT \tag{7.86}$$

which may be written

$$\overline{\text{displacement} \times \text{opposing force}} = kT \tag{7.87}$$

In the special case that E' is proportional to x^2 the simpler result

$$\bar{E}' = \tfrac{1}{2}kT \tag{7.88}$$

is again obtained.

The equipartition theorem is obviously a powerful tool, particularly since it can be applied to macroscopic bodies, but is restricted to situations which may be described by the classical limit of quantum mechanics. The use of the equipartition theorem to obtain the mean square fluctuation about the mean energy of a system in thermal equilibrium will be discussed in the next section.

7.7 Fluctuations about equilibrium

The general result for the mean square fluctuations of a variable x was given in eqn 7.12

$$\overline{(x - \bar{x})^2} = \overline{x^2} - (\bar{x})^2$$

The value of $\overline{x^2}$ can often be found using the equipartition theorem and is equal to the mean square fluctuation of the system if $\bar{x}$ is zero. The mean square angular rotation of the galvanometer suspension with torque constant c for example (eqn 7.85) is given by

$$\tfrac{1}{2}c\overline{\theta^2} = \tfrac{1}{2}kT$$

The mean angle $\bar{\theta}$ is obviously zero since the suspension is equally likely to turn clockwise or anticlockwise from the equilibrium position so

$$\sqrt{(\theta - \bar{\theta})^2} = \sqrt{\frac{kT}{c}}. \tag{7.89}$$

A suspension at temperature T will rotate randomly about the equilibrium position such that the root mean square amplitude of the angular displacement is given by eqn 7.89. This fluctuation in the angle θ will mask the rotation due to a small steady current I. The rotation due to the current is proportional to $c\theta$ so the minimum observable current is given by

$$I_{\min} \propto \sqrt{(ckT)}. \tag{7.90}$$

The thermal fluctuations (noise) can only be reduced for a given suspension by reducing the temperature of the system.

The thermal noise present in an electronic amplifier may be analysed in a similar fashion to the galvanometer suspension and is seen to provide a fundamental limitation to the size of signal that can be detected. The amplifiers used in radioastronomy and other subjects where the signals are extremely weak are sometimes immersed in liquid helium (4.2 K) to achieve a theoretical improvement in performance of $\sqrt{(300/4.2)}$ compared to the same amplifier at room temperature. There are many other problems apart from thermal noise in the design of amplifiers to detect small signals, but thermal noise is the final limitation on the performance of the system.

The mean square fluctuation in the velocity of a gas molecule in a classical gas at temperature T is given immediately by eqn 7.12 since the mean velocity is zero (remember that velocity involves magnitude and direction and there can be no preferred direction in the gas in thermal equilibrium) so

$$\overline{(v - \bar{v})^2} = \overline{v^2} = \frac{3kT}{m}. \tag{7.91}$$

The mean square fluctuation in the velocity increases with temperature and is inversely proportional to the mass of the particle.

The equipartition theorem and the general eqn 7.12 can therefore be used to obtain useful information about the mean square fluctuations about equilibrium of any system which is in the classical regime and has a mean value of zero for the property under consideration. The *distribution* of fluctuations however is not characterised by eqn 7.12 alone and a complete theory for a particular system is required when more than just the mean square value of the fluctuation about the equilibrium value is required. The Maxwell velocity distribution for the molecules of a classical perfect gas has a Gaussian form (eqn 6.1 and Chapter 8) with mean value zero and mean square deviation given by eqn 7.91 but the Gaussian form could not be deduced from eqn 7.12.

The general equation for the mean square fluctuation of the thermal energy of any system in thermal equilibrium at temperature T (eqn 7.18)

$$\overline{(E - \bar{E})^2} = kT^2 C_V$$

has already been discussed in Section 7.1. The ratio of the root mean square fluctuation of the energy to the mean energy was seen to be of order $1/\sqrt{N}$ in the region where the equipartition theorem could be applied.

Fluctuations are of importance however when C_V becomes very large, as at the critical point for a phase transition, such as liquid to vapour (Section 5.4). The thermodynamic result (eqn 5.11)

$$C_P - C_V = - T \left(\frac{\partial P}{\partial T} \right)_V^2 \bigg/ \left(\frac{\partial P}{\partial V} \right)_T$$

and the condition (eqn 2.13)

$$\left(\frac{\partial P}{\partial V} \right)_T = 0 \tag{7.92}$$

at the phase transition show that $(C_P - C_V)$ goes to infinity at the critical point and from experiment it appears that C_V also approaches infinity, although more weakly than C_P (Section 10.4). The isothermal compressibility of the system goes to infinity at the critical point (eqn 7.92) and large fluctuations occur in the density of the system. This is to be expected since, for example, as the gas is cooled droplets of liquid will begin to form in the vapour. The grand canonical ensemble may be used to prove that the mean square fluctuation in the number of particles in volume V is given by

$$\overline{(N - \bar{N})^2} = \frac{-kT(\bar{N})^2}{V^2 \left(\dfrac{\partial P}{\partial V} \right)_T} \tag{7.93}$$

(see appendix at end of chapter (eqn 7.116).)
The importance of fluctuations is given by

$$\sqrt{\left[\frac{\overline{(N - \bar{N})^2}}{(\bar{N})^2} \right]} = \sqrt{\frac{-kT}{V^2 \left(\dfrac{\partial P}{\partial V} \right)_T}}. \tag{7.94}$$

The right-hand side is simply equal to $1/\sqrt{N}$ for a perfect gas. Fluctuations in the number density are therefore unimportant (except at the critical point when $(\partial P/\partial V)_T$ goes to zero) unless the volume V under consideration is very small.

7.8 Negative temperature

The thermodynamic temperature (T) was introduced in Section 3.1 as the integrating factor for the imperfect differential dQ_R such that the perfect differential dS was defined by the equation

$$dS = \frac{dQ_R}{T}.$$

The thermodynamic temperature may then be written

$$\frac{1}{T} = \left(\frac{\partial S}{\partial U}\right)_V$$

and has been treated as an inherently positive quantity up to this point.

The temperature also appears as a positive quantity in statistical mechanics since the partition function is written

$$Z = \sum_r e^{-\beta E_r}$$

where $\beta = 1/kT$. There is no upper limit to the allowed energy states of a system such as the simple harmonic oscillator and therefore β must remain positive if the partition function is to be finite.

The possibility of negative temperatures can only exist for systems with an upper allowed energy level of finite energy for which the partition function would remain finite if β became less than zero. In the strictest sense it is easy to see that no such system can exist and that in true thermodynamic equilibrium the thermodynamic temperature must be positive but it will be shown that it is sometimes possible to produce a state in one aspect of a system which may be defined by a negative temperature.

The theory of the thermal capacity of solids, discussed in Section 10.3, assumes that the energy of a crystal may be written as the sum of a number of independent terms, as for example

$$E = E_1 + E_e + E_s + \dots \tag{7.95}$$

where E_1 is the energy due to lattice vibrations, E_e is the energy of free electrons and E_s is the Schottky energy due to the splitting of energy levels. The lattice energy is proportional to the temperature in the high-temperature limit and in this sense there is no upper limit to the allowed energy of the crystal. (Above the melting point of course the crystal ceases to exist but the energy of a liquid and a gas also have no upper limit.) In thermodynamic equilibrium all the terms on the right-hand side of eqn 7.95 must be described by the same temperature and since the lattice energy is restricted to positive temperatures only positive temperature are possible, in agreement with the assumptions made earlier in the book.

A more general definition of temperature however does admit the possibility of negative temperatures. The two-level system for spin 1/2 in a magnetic field, Fig. 7.4, [or in general the $(2I + 1)$−level system for spin I] satisfies the condition that there exists a finite upper energy level. The nuclear spins in certain insulating crystals (for example LiF) are found to establish internal equilibrium in times of the order of 50 μs (the spin–spin relaxation time) but to require some 300 s to come into equilibrium with the rest of the crystal (the spin–lattice relaxation time). It is therefore legitimate to treat the nuclear spins as isolated from the lattice and in internal equilibrium with a spin temperature (T_s) for times long compared to 50 μs and short compared to 300 s.

A spin temperature T_s may be defined if the probability of occupation of a spin state of energy E_r is given by

$$P_r = \frac{e^{-\beta_s E_r}}{\sum_r e^{-\beta_s E_r}}$$

where as usual $\beta_s = 1/kT_s$. This condition is always satisfied for the two-level system and the equation for the number of spins in the upper state (separated by an energy Δ from the ground state)

$$N_1 = \frac{N}{1 + e^{\beta_s \Delta}} \tag{7.96}$$

may be thought of as defining β_s. The number of spins in the upper state tends to $N/2$ as T_s goes to plus infinity $(\beta \to 0+)$. If N_1 can be made greater than $N/2$ the system must be described according to eqn 7.96 by a negative value of β_s.

The experimental method for attaining negative spin temperatures is to place the sample in a large steady magnetic field $\mu_0 H^*$. Then

$$\Delta = 2\mu\mu_0 H^*$$

where μ is the nuclear moment (Section 7.4). The spin temperature is then equal to the lattice temperature (T_L) and N_1 $(<N/2)$ is given by eqn 7.96 with β_s equal to β_L. The spins cannot follow a sufficiently rapid reversal of the field $\mu_0 H^*$ to $-\mu_0 H^*$ which then leaves $N_1 > N/2$ and the spin temperature equal to $-T_L$.

The population of the upper state given in eqn 7.96 may be written in the symmetrical form

$$N_1 = \frac{N}{2} + \frac{N(1 - e^{\beta_s \Delta})}{2(1 + e^{\beta_s \Delta})} \tag{7.97}$$

since a change from β_s to $-\beta_s$ changes the sign of the second term (Fig. 7.7). Notice that the equal population of the two levels corresponds to β_s equal to zero, that is to say to T_s equal to *both* plus and minus infinity. Negative

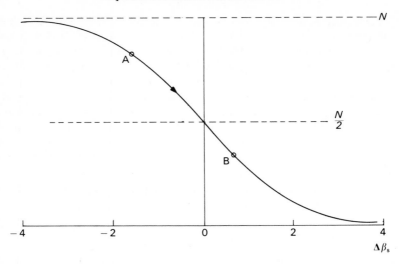

Fig. 7.7 The temperature of a two-level system may be defined in terms of the population of the upper level if the spin–spin relaxation time is much shorter than the spin–lattice relaxation time. A system with negative temperature (A) passes to positive temperature (B) via β_s equal to zero ($T_s = \pm \infty$) not via $T = 0$ K.

temperatures correspond to higher energies ($E_s > N\Delta/2$) than positive temperatures and it is logical to describe them as 'hotter' than any positive temperature. The third law of thermodynamics is unaffected by the possibility of negative temperature because the spin system returns to the lattice temperature via an infinite temperature ($N = N/2$) not via absolute zero (Fig. 7.7).

The spin system can only be brought into the region of negative temperature by a rapid reversal of the magnetic field. A quasistatic change of the field would leave the spins with the temperature of the lattice since the time involved would be greater than the spin–lattice relaxation time. A Carnot cycle cannot therefore be constructed to work between reservoirs with temperature of opposite sign.

A spin system at positive temperature will tend to absorb an alternating magnetic field at frequency (ν_0) such that

$$h\nu_0 = \Delta = 2\mu\mu_0 H^* \tag{7.98}$$

since the probability of a spin being excited by the field to the upper level by the absorption of a photon is equal to the probability that a spin in the upper state will be induced to emit a photon (stimulated emission) and drop to the lower level. The power absorbed is therefore proportional to the difference in the populations of the levels (for a given system and alternating field)

$$P \propto (N_0 - N_1). \tag{7.99}$$

A system with a negative temperature $(N_1 > N_0)$ has a negative power absorption, that is to say amplifies the signal at frequency v_0, but this response will only continue until the populations of the two levels become equal.

An amplifier which can operate continuously may be constructed if some method is found of repopulating the higher energy level. A maser (*M*icrowave *A*mplification by *S*timulated *E*mission of *R*adiation) depends upon this principle. One type of three-level maser is illustrated in Fig. 7.8. The energy separation (Δ) of the levels 1 and 2 is arranged to satisfy eqn 7.98 for the signal frequency v_0 by adjusting the magnetic field $\mu_0 H^*$. The upper state 3 is now populated by an alternating magnetic field at the pump frequency (v_p) given by

$$hv_p = \Delta_{13} \tag{7.100}$$

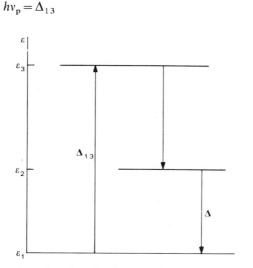

Fig. 7.8 In one version of a three-level maser the energy state 2 is populated by exciting the state 3 with a pumping oscillator. A signal with frequency given by eqn 7.98 is then amplified.

where Δ_{13} is the energy separation between the states. The population of state 3 can be raised to half that of the equilibrium populations of states 1 and 3 by a sufficiently strong pumping field. The population of state 2 depends upon the rate at which it is populated from state 3 and the rate at which transitions are made from 2 back to the ground state. A sufficiently fast decay from state 3 to state 2 will leave N_2 greater than N_1 and the signal at v_0 will be amplified. The states 1 and 2 are sometimes described by a negative temperature but it is clear that this is no more than an expression of the relative populations of the two states and has little relation to the spin temperature discussed earlier which corresponded to internal equilibrium between the spins.

The maser is used in the microwave region (3–30 GHz) as a low-noise communications amplifier. The required energy separation of states 1 and 2 at these frequencies (10^{-5}–10^{-4} eV) is obtained by applying a magnetic field to a system of electron spins rather than the nuclear spins discussed earlier in this section. A low concetration of Cr^{3+} ions in alumina (ruby) has been found to be a suitable spin system. The ruby maser is usually operated at 4 K since the noise in the system is mainly thermal noise which is proportional to the absolute temperature. The other contribution to the noise is due to spontaneous emission from state 2 to the ground state but at microwave frequencies this term is small relative to thermal noise. However the laser (*Light Amplification by Stimulated Emission of Radiation*), which operates on the same principle as the maser but at much higher frequencies, is a noisy amplifier since the noise due to spontaneous emission is now much greater than thermal noise.

7.9 Conclusion

The problem of the thermodynamic fundamental equation discussed in Part 1 has been shown in this chapter to be equivalent, for a system with a fixed number of particles, to the evaluation of the partition function of the system. The fundamental equation may then be written in terms of the Helmholtz free energy and the partition function of the system

$$F = -kT \ln Z \tag{7.101}$$

$$Z = \sum_r e^{-\beta E_r}. \tag{7.102}$$

The partition function is a function of temperature (since $\beta = 1/kT$) and of the volume of the system if the states r depend upon the volume, so the right-hand side of eqn 7.102 is a function of (T, V) as was required if the Helmholtz free energy was to define the fundamental equation of the system.

Statistical mechanics goes beyond thermodynamics however in that it provides a means for the study of the fluctuations of the properties of the system about their equilibrium values. When the fluctuations are small (as is usually the case for large systems) the mean value of the energy of the system has been seen to be well defined and may be treated as the thermodynamic internal energy discussed in Part I.

The original statistical definition of the entropy due to Boltzmann, eqn 6.7, was seen to be correctly extensive but did not prohibit a decrease in the entropy of an isolated system in thermal equilibrium. The probability of a significant (macroscopically observable) decrease in the entropy of a large system was however found to be vanishingly small.

The Gibbs definition of the entropy

$$S = -k\Sigma P_r \ln P_r \tag{7.103}$$

avoids this difficulty of principle inherent in the Boltzmann entropy since the ensemble probabilities (P_r) are independent of time for a system in thermal equilibrium. The entropy defined by eqn 7.103 cannot therefore decrease.

The third law of thermodynamics in the form (eqn 4.22)

$$S_{\to S_0} \qquad T_{\to 0\,\mathrm{K}}$$

where S_0 is a constant, is equivalent to the statement that the sum on the right-hand side of eqn 7.103 becomes independent of temperature at sufficiently low temperature. If the lowest energy level of the system is g-fold degenerate then

$$S_0 = k \ln g \qquad\qquad\qquad (7.104)$$

which is certainly negligible by macroscopic standards ($\approx Nk$). As was remarked earlier it is thought that the true ground state of any system is non-degenerate ($g = 1$) and then S_0 is equal to zero, which was the strongest statement of the third law given in Chapter 4. This simple formulation of the third law is not however really satisfacotry. The thermal energy required to excite a particle in a box from the ground to the first excited energy state is only of the order of 10^{-14} K but the concept of the entropy becoming a constant at 'low temperature' was derived from experiments at temperatures near 4 K. There are therefore many particles in excited states at even the lowest temperature accessible to experiment.

The explanation of this discrepancy between the observed temperature at which the entropy becomes independent of temperature and that expected if the system was required to be in its ground state, is once again related to the fact that the degeneracy factor g in equation 7.104 may be very large by normal standards (say 10^{10}) without making a significant contribution to the entropy of the system (eqn 6.20). In terms of the density of states function, eqn 7.104 may be written

$$S = k \ln D(E)\,\mathrm{d}E. \qquad\qquad\qquad (7.105)$$

Casimir first pointed out that provided $D(E)\,\mathrm{d}E$ increases sufficiently slowly with energy the contribution to S_0 will be immeasurably small at experimental 'low temperature' even though many states of the system are occupied. The third law of thermodynamics should therefore be understood in this rather more restricted way than the simple statement involving the true ground state of the system which is usually not accessible to experiment.

The theoretical framework of statistical mechanics and thermodynamics for systems with a fixed number of particles in thermal equilibrium has now been taken as far as is necessary for the purpose of this book. In the next two chapters, the original version of statistical mechanics (the kinetic theory of the classical perfect gas) will be discussed for both equilibrium and non-equilibrium states. A number of further examples of the application of the theory developed so far will be found in the last part of the book.

*7.10 Appendix. Systems with a variable number of particles

The concept of ensemble averaging may be extended to a system containing a variable number of particles. (The thermodynamics of such systems was considered in Section 4.5). The system of interest of volume V and at temperature T is now replaced for the purposes of calculation by an ensemble of replicas, as in Section 7.1, but the replicas are now free to exchange particles. It is clear that, since both energy and particles can be transferred between the replicas, only the mean values of the energy and the number of particles are now constant. However in the canonical ensemble it was found that the energy fluctuations about $\bar{E}$ for a macroscopic system are usually unimportant and similarly in the new ensemble, the grand canonical ensemble, fluctuations about both $\bar{E}$ and $\bar{N}$ are usually unimportant.

The definition of the entropy, eqn 7.30,

$$S = -k \sum_r P_r \ln P_r \tag{7.106}$$

may be extended to a system with a variable number of particles but it is necessary to remember that in general the states of the system will be a function of the number of particles in the volume V. (As the concentration of particles at a given temperature increases a system might, for example, go from a perfect gas to a dense gas and then to a liquid.)

Equation 7.106 must therefore be written in terms of P_{Nr}, where P_{Nr} is the probability of the state r in a system containing N particles. On introducing Lagrange undetermined multipliers as in Section 7.3, and maximizing the entropy under the conditions that the mean number of particles and the mean energy of the system is constant.

$$P_{Nr} = \frac{e^{\beta(\mu N - E_{Nr})}}{\mathscr{Z}} \tag{7.107}$$

$$\mathscr{Z} = \sum_N Z_N e^{\beta \mu N} \tag{7.108}$$

$$Z_N = \sum_r e^{-\beta E_{Nr}}. \tag{7.109}$$

Note that the above equations reduce to the canonical ensemble results if N is taken to be a constant. The function β is therefore simply $1/kT$ as before and μ will be shown to be the chemical potential introduced in Section 4.5.

On substituing equation 7.107 into equation 7.106

$$TS = \bar{E} - \mu\bar{N} + kT \ln \mathscr{Z}$$

and taking mean values to correspond to thermodynamic quantities the equation

$$\Omega = -kT \ln \mathscr{Z}(T, V, \mu) \tag{7.110}$$

---

*This section may be omitted on a first reading.

follows from the thermodynamic identity, eqn 4.34

$$\Omega = -PV = U - TS - \mu N$$

Equation 7.110 is the fundamental equation for a system with a given volume and temperature but a variable number of particles, Ω is called the grand potential and $\mathscr{Z}$ the grand partition function. All the properties of the system may be calculated if $\mathscr{Z}$ is known as a function of (T, V, μ) using the results of Section 4.5,

$$P = -\left(\frac{\partial \Omega}{\partial V}\right)_{T,\mu} \tag{7.111}$$

$$\bar{N} = -\left(\frac{\partial \Omega}{\partial \mu}\right)_{T,V}. \tag{7.112}$$

The mean square fluctuation of the number of particles may be found by using the same approach as in Section 7.7 for the fluctuation in the energy of a system in the canonical ensemble. Using eqn 7.107 and the definition of a mean value,

$$\bar{N} = \sum_{N,r} N P_{Nr} = \frac{kT}{\mathscr{Z}}\left(\frac{\partial \mathscr{Z}}{\partial \mu}\right)_{T,V} \tag{7.113}$$

$$\overline{N^2} = \sum_{N,r} N^2 P_{Nr} = \frac{(kT)^2}{\mathscr{Z}}\left(\frac{\partial^2 \mathscr{Z}}{\partial \mu^2}\right)_{T,V} \tag{7.114}$$

$$\left(\frac{\partial \bar{N}}{\partial \mu}\right)_{T,V} = \frac{kT}{\mathscr{Z}}\left(\frac{\partial^2 \mathscr{Z}}{\partial \mu^2}\right)_{T,V} - \frac{kT}{\mathscr{Z}^2}\left(\frac{\partial \mathscr{Z}}{\partial \mu}\right)^2_{T,V} \tag{7.115}$$

$$\therefore \ \overline{(N-\bar{N})^2} = \overline{N^2} - (\bar{N})^2 = kT\left(\frac{\partial \bar{N}}{\partial \mu}\right)_{T,V}$$

The right-hand side of 7.115 may be manipulated by thermodynamics into a more useful form since from Exercise 4.13,

$$d\mu = -s\,dT + v\,dP$$

where s is the entropy, and v the volume, per particle. Hence,

$$\left(\frac{\partial \mu}{\partial v}\right)_T = v\left(\frac{\partial P}{\partial v}\right)_T.$$

In a uniform system containing N particles the whole volume of the system V is equal to vN and

$$dv = -\frac{V}{N^2}\,dN + \frac{1}{N}\,dV$$

$$\therefore \ -\frac{N^2}{V}\left(\frac{\partial \mu}{\partial N}\right)_{T,V} = V\left(\frac{\partial P}{\partial V}\right)_{T,N}.$$

Identifying N in a thermodynamic equation with $\bar{N}$ in statistical mechanics as usual it follows that

$$\frac{\overline{(N-\bar{N})^2}}{(\bar{N})^2} = -\frac{kT}{V^2}\left(\frac{\partial V}{\partial P}\right)_{T,N} = \frac{kT\kappa_T}{V}. \tag{7.116}$$

When the perfect gas law holds the right hand side is simply equal to $1/N$ so the fluctuation in $\bar{N}$ is of order $1/\sqrt{N}$ as for the energy, eqn. 7.19, and $\bar{N}$ is well defined. As the gas approaches the critical point the isothermal compressibility tends to infinity, eqn 2.13, and the fluctuations in the density of the system become very large, see Section 7.7.

In general the solution of eqn 7.110

$$\Omega = -kT \ln \mathscr{L}(T, V, \mu)$$

is a N body problem, as was the canonical result

$$F = -kT \ln Z(T, V).$$

The canonical partition function however was seen to factorize, eqn 7.27,

$$Z = Z^N$$

for a weakly coupled system of localized particles and to, eqn 7.28,

$$Z = \frac{Z^N}{N!} \tag{7.117}$$

for a perfect gas in the classical limit. The quantum gas can not however be expressed in terms of single particle canonical partition functions because of the restraint that N be held constant. This restriction is removed in the grand canonical ensemble so it is permissible to write,

$$\Omega = \sum_r \Omega_r; \quad \mathscr{L} = \prod_r \mathscr{L}_r$$

where $\Omega_r, \mathscr{L}_r$ refer to a single particle state r. The other states of the system may be thought of as providing the particle reservoir for state r.

In the case of a Fermi–Dirac gas for example the occupation number of the single particle state can only be 0 or 1 so

$$\mathscr{L}_r = 1 + e^{\beta(\mu - \varepsilon_r)}$$
$$\bar{n}_r = -\left(\frac{\partial \Omega_r}{\partial \mu}\right)_{T,V} = kT\left(\frac{\partial \ln}{\partial \mu}\mathscr{L}_r\right)_{T,V} = \frac{1}{e^{\beta(\varepsilon_r - \mu)} + 1} \tag{7.118}$$

this is the exact result for the thermal average number of particles in the single particle state r and is identical to the approximate result for the most probable number of particles given in Chapter 6. The calculations for bosons and the classical limit are given as exercises at the end of Chapter 7.

Exercises

7.1. Calculate the value of Δ/kT for a nucleus of spin $\frac{1}{2}$ with magnetic moment 5×10^{-27} JT^{-1} in a field of 1 T(10 kG) at a temperature of 1 K and repeat the calculation for an ion with magnetic moment 9×10^{-24} JT^{-1}.

7.2. Show that in the high-temperature limit the heat capacity of a two-level system is proportional to the inverse square of the temperature.

7.3. Find an expression for the entropy of a system of one-dimensional simple harmonic oscillators as a function of temperature.

7.4. Show that a gas of particles moving sufficiently close to the speed of light (c) to have the energy related to the momentum by $\varepsilon = cp$ obeys the equation $PV = \bar{E}/3$.

7.5. The temperature of the chromosphere of the sun is approximately 5×10^3 K. Find the relative number of hydrogen atoms in the energy levels with $n = 1, 2, 3, 4$. The Rydberg constant is 1.1×10^7 m^{-1}.

7.6. The moment of inertia of a molecule is 4.6×10^{-48} kg m^2. Find the relative populations of the $J = 0, 1, 2, 3, 4$ rotational energy levels at 300 K. At what temperature are the $J = 2$ and $J = 3$ levels equally populated?

7.7. Express the Gibbs free energy and the enthalpy as functions of: (a) the partition function of the whole system; (b) for N weakly coupled localized particles; and (c) for N particles of a gas in the classical limit.

7.8. Given the mean energy and the entropy in the form

$$\bar{E} = \sum_r P_r E_r, \quad S = -k \sum_r P_r \ln P_r, \quad P_r = Ce^{-\beta E_r},$$

and that, from quantum mechanics, an infinitely slow reversible adiabatic change of volume does not cause transitions between states, show: (a) under such adiabatic conditions

$$d\bar{E} = \sum_r P_r dE_r = dW_R$$

i.e. $dW_R = \left(\dfrac{\partial \bar{E}}{\partial V}\right)_s dV$; (b) in a general reversible process $dQ_R = \sum_r E_r dP_r$; and (c) $dS = k\beta dQ_R$ i.e. $\beta = 1/kT$.

7.9. Find expressions, in both the general case and when the tension is small, for: (a) the coefficient of thermal expansion; (b) the heat capacity at constant length; and (c) the heat capacity at constant tension for the model of a rubber fibre given in Section 7.4.

7.10. Derive eqns 7.57 and 7.59.

7.11. A moving coil galvanometer consists of a coil, in a uniform magnetic field B_0, suspended from a fibre of torque constant c. A current i in the coil produces a deflection

$$\theta = nAB_0 i/c$$

where n is the number of turns and A the area of the coil. The period of oscillation of a coil of moment of inertia I is given by

$$\tau = 2\pi (I/c)^{\frac{1}{2}}$$

The resistance of the coil and the circuit in which it is placed is adjusted to give critical damping

$$R_c = n^2 A^2 B_0^2 / 2(Ic)^{\frac{1}{2}}.$$

Assume that the minimum detectable current produces a deflection equal to the thermal r.m.s. value of θ and hence show that

$$i_{min} = (\pi k T / R_c \tau)^{\frac{1}{2}}.$$

How does i_{min} depend upon (a) the temperature of the coil, (b) the time taken over a measurement? (These results are also valid for electronic circuits).

7.12. Write down the expression for the mean energy of a system in the canonical ensemble when the energy states may be treated as a continuous function. Hence show that a negative temperature is only possible if the density of states decreases at least exponentially with energy.

7.13. A single particle of a weakly interacting system can only take on energies 0 or Δ. Two such systems, each containing N particles, are initially at temperature T and T/α. Find an expression for the final temperature when they are joined together and evaluate it for α equal to 1 and -1.

7.14. Show that for bosons the grand partition function for the single particle state r is given by,

$$\mathscr{L}_r = [1 - e^{\beta(\mu - \varepsilon_r)}]^{-1}$$

and hence

$$\bar{n}_r = [e^{\beta(\varepsilon_r - \mu)} - 1]^{-1}$$

provided $\mu < \varepsilon_0$. [This problem may be omitted on a first reading. A knowledge of the material in Section 7.10 is required.]

7.15. Taking the canonical partition function for a perfect gas in the classical limit from equation A6.12 show that

$$\mathscr{Z} = \exp(ze^{\beta\mu})$$

$$\Omega = -kTze^{\beta\mu}$$

$$PV = \bar{N}kT.$$

(Hint: $e^x = \sum_n \frac{1}{n!} x^n$).

PART III
Kinetic theory

8

Kinetic theory of gases—I

The kinetic theory of the classical perfect gas (largely due to Maxwell and Boltzmann) was the first statistical theory to be introduced into physics. The equilibrium results of kinetic theory can now all be obtained as special cases of the general results of statistical mechanics obtained in Chapter 7 but the detailed nature of the kinetic theory can give added insight into processes such as the pressure exerted by a gas on the wall of a container. The kinetic theory approach can also be extended to non-equilibrium situations and, as shown in the next chapter, may be used to calculate the transport coefficients (thermal conductivity, coefficient of viscosity, etc.) of a perfect gas. The kinetic theory of the perfect gas is of great importance in applied physics because the results are directly relevant to real gases at low pressure as in vacuum systems, the upper atmosphere, and space research.

The model of a gas on which the kinetic theory is based is that of a set of molecules moving rapidly at random throughout the whole volume of the containing vessel. The energy of a monatomic gas is set equal to the sum of the kinetic energies of the molecules. In the simplest model the molecules are treated as elastic spheres (this assumption is discussed in more detail in Chapter 9) which on average travel a distance (the mean free path) which is large compared to their diameter, before making a collision. The total volume occupied by the molecules is assumed to be much less than the volume of the containing vessel. The kinetic theory will be seen in Chapter 9 to be internally consistent in that the mean free path in a gas at NTP calculated using the measured and predicted transport coefficients is indeed much greater than the molecular diameter. (Typical values are given in Appendix VII.)

The statistical nature of the kinetic theory should be clear from the discussion of Chapter 7. An individual molecule will change its velocity after each collision. Only the probability of the molecule being found to have a given velocity (or small range of velocities) can be specified, but for a gas in thermal equilibrium the fraction of the molecules within the given range of velocities remains constant. The equilibrium results of kinetic theory will be discussed in this chapter and the transport theory in Chapter 9.

8.1 Distribution functions

A number of distribution functions are sometimes useful in kinetic theory calculations although they do not contain any information that cannot be found from the partition function. The distribution function for some property x is defined either by the relation

$$f(x)\mathrm{d}x = \text{probability of } x \text{ lying in the range } x \text{ to } x+\mathrm{d}x$$

or alternatively

$$F(x)\,\mathrm{d}x = \text{number of molecules with } x \text{ in the range } x \text{ to } x+\mathrm{d}x.$$

In the first case the integral of the distribution function over all values of x is equal to unity and in the second case is equal to the total number of molecules (N). The two distribution functions are simply related

$$F(x) = Nf(x).$$

It is usual in kinetic theory however to write simply $f(x)$ for either distribution function. The parameter x may be either a vector (for example momentum, velocity) or a scalar (for example energy, speed) quantity.

The fundamental distribution function may be derived directly from the classical partition function of a single particle

$$f(\boldsymbol{p}, \boldsymbol{r})\mathrm{d}^3\boldsymbol{p}\,\mathrm{d}^3\boldsymbol{r} = \frac{\mathrm{e}^{-\beta\varepsilon}\mathrm{d}^3\boldsymbol{p}\,\mathrm{d}^3\boldsymbol{r}}{\displaystyle\int \mathrm{e}^{-\beta\varepsilon}\mathrm{d}^3\boldsymbol{p}\,\mathrm{d}^3\boldsymbol{r}} \tag{8.1}$$

$$= \text{probability that a molecule has momentum in the range } \boldsymbol{p}$$
$$\text{to } \boldsymbol{p}+\mathrm{d}\boldsymbol{p} \text{ and position in the range } \boldsymbol{r} \text{ to } \boldsymbol{r}+\mathrm{d}\boldsymbol{r}.$$

The limits of the integrals are taken to be plus and minus infinity for the momentum and the whole of the volume of the container for the position coordinates. The extension to plus and minus infinity of the momentum variable is not of course consistent with the theory of relativity but this is of no practical importance since the exponential term is negligible at such high energies. When the momentum of a particle is independent of its position the left-hand side of eqn 8.1 may be factorized

$$f(\boldsymbol{p}, \boldsymbol{r})\mathrm{d}^3\boldsymbol{p}\,\mathrm{d}^3\boldsymbol{r} = f(\boldsymbol{p})\mathrm{d}^3\boldsymbol{p}f(\boldsymbol{r})\mathrm{d}^3\boldsymbol{r}$$

and when gravitational effects are neglected the energy of a particle is independent of its position so eqn 8.1 becomes

$$f(\boldsymbol{p})\mathrm{d}^3\boldsymbol{p} = \frac{\mathrm{e}^{-\beta\varepsilon}\mathrm{d}^3\boldsymbol{p}}{\displaystyle\int \mathrm{e}^{-\beta\varepsilon}\mathrm{d}^3\boldsymbol{p}}. \tag{8.2}$$

The energy of a particle is related to the momentum by

$$\varepsilon = \tfrac{1}{2}mv^2 = \frac{p^2}{2m} = \frac{1}{2m}(p_x^2 + p_y^2 + p_z^2)$$

The normalized velocity distribution is most easily obtained by noticing that the triple integral in eqn 8.2 is separable in cartesian coordinates.

$$f(\mathbf{p})\mathrm{d}^3\mathbf{p} = f(\mathbf{p}_x)\mathrm{d}\mathbf{p}_x f(\mathbf{p}_y)\mathrm{d}\mathbf{p}_y f(\mathbf{p}_z)\mathrm{d}\mathbf{p}_z .$$

Therefore

$$f(\mathbf{p}_x)\mathrm{d}\mathbf{p}_x = \frac{e^{-\frac{\beta p_x^2}{2m}}\mathrm{d}\mathbf{p}_x}{\displaystyle\int_{-\infty}^{\infty} e^{-\frac{\beta p_x^2}{2m}}\mathrm{d}\mathbf{p}_x} .$$

The denominator is a standard definite integral (Appendix II) so

$$f(\mathbf{p}_x)\mathrm{d}\mathbf{p}_x = \left(\frac{1}{2\pi mkT}\right)^{\frac{1}{2}} e^{-\frac{\beta p_x^2}{2m}}\mathrm{d}\mathbf{p}_x \qquad (8.3)$$

$$f(\mathbf{p})\mathrm{d}^3\mathbf{p} = \left(\frac{1}{2\pi mkT}\right)^{\frac{3}{2}} e^{-\frac{\beta p^2}{2m}}\mathrm{d}^3\mathbf{p} \qquad (8.4)$$

are the normalized momentum distribution functions in one and three dimensions.

The velocity distribution function follows immediately since

$$\mathbf{p} = m\mathbf{v} \qquad \mathbf{p}_x = m\mathbf{v}_x$$

$$f(\mathbf{v}_x)\mathrm{d}\mathbf{v}_x = f(\mathbf{p}_x)\left(\frac{\mathrm{d}p_x}{\mathrm{d}v_x}\right)\mathrm{d}\mathbf{v}_x .$$

Therefore

$$f(\mathbf{v}_x)\mathrm{d}\mathbf{v}_x = \left(\frac{m}{2\pi kT}\right)^{\frac{1}{2}} e^{-\frac{mv_x^2}{2kT}}\mathrm{d}\mathbf{v}_x \qquad (8.5)$$

$$f(\mathbf{v})\mathrm{d}^3\mathbf{v} = \left(\frac{m}{2\pi kT}\right)^{\frac{3}{2}} e^{-\frac{mv^2}{2kT}}\mathrm{d}^3\mathbf{v} . \qquad (8.6)$$

The momentum and velocity distribution functions therefore have maximum values at the origin and are symmetric about the origin since there can be no preferred direction in space for a system in thermal quilibrium when gravitational effects may be neglected. The one-dimensional velocity distribution function is shown in Fig. 8.1. The area under the curve is always equal to unity but the width of the distribution at half-height increases and the maximum value decreases as the temperature is raised.

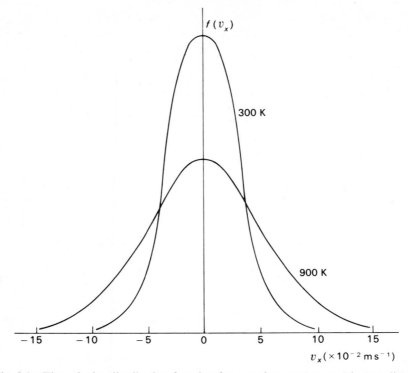

Fig. 8.1   The velocity distribution function for a perfect gas (oxygen) in one dimension for temperature 300 K and 900 K. The area under the curve is always equal to unity. The mean value of the velocity is zero.

Two features of the velocity distribution function are of some interest in connection with the nature of the quantity defined as 'heat' in Part I. It is clear from Fig. 8.1 that the average velocity (magnitude and direction) of the molecule relative to the walls of the containing vessel is zero. The mean square velocity however is not zero (Section 8.2) and increases with the temperature. The thermal energy is therefore associated with the *random* motion of the molecules within the container but external work involves the movement of a piston in a given direction. The conversion of heat into work therefore involves a change from the random motion of many particles to the ordered motion of the piston so it is not surprising that a thermal cycle cannot operate with an efficiency of unity.

The second feature of the velocity distribution function which should be noted is that the width of the distribution increases with temperature. The probability of finding the molecule in a given small range of velocity therefore decreases as the temperature increases. An alternative way of expressing this is to say that our *information* about the velocity of the molecule has decreased as the temperature increased and since from eqn 7.70 an increase of temperature at constant volume corresponds to an increase in the entropy of the gas there must be a relationship between information and (the negative of) the entropy. This point will not be pursued further here but the connection between information and entropy is sometimes used as a basis for statistical mechanics using eqn 7.30

$$S = -\sum_r P_r \ln P_r$$

to define the entropy of any system whether or not it is in thermal equilibrium.

The speed distribution function is found by integrating the velocity distribution function over all directions at constant $|v|$. The integration is most simply carried out by using polar coordinates.

$$d^3v = v^2 \, dv \sin\theta \, d\theta \, d\phi$$

$$f(v)dv = \left(\frac{m}{2\pi kT}\right)^{\frac{3}{2}} v^2 \, dv \int_0^\pi \sin\theta \, d\theta \int_0^{2\pi} d\phi$$

$$= 4\pi v^2 \left(\frac{m}{2\pi kT}\right)^{\frac{3}{2}} e^{-\frac{mv^2}{2kT}} dv \tag{8.7}$$

The speed distribution function is shown in Fig. 8.2. The most probable speed is not zero but may be found by differentiating eqn 8.7

$$v_{\text{mp}} = \sqrt{\frac{2kT}{m}}$$

The energy distribution function was derived in Chapter 7 but may also be found from eqn 8.7 using

$$\varepsilon = \tfrac{1}{2} mv^2 \quad d\varepsilon = mv \, dv$$

$$f(\varepsilon)d\varepsilon = 2\left(\frac{\varepsilon}{\pi k^3 T^3}\right)^{\frac{1}{2}} e^{-\frac{\varepsilon}{kT}} d\varepsilon. \tag{8.8}$$

The energy distribution function is shown in Fig. 8.3. The remarkable feature of eqn 8.8 is that the energy distribution function is independent of the particle mass. The energy distribution function for all perfect gases is the same at temperature T. This result goes beyond the earlier statement of the equipartition theorem, that the mean energy of a particle is independent of the mass, to show that the distribution of the energy is also independent of the mass. In principle the perfect gas temperature T could therefore be defined by measuring the distribution function and using eqn 8.8 to find the temperature.

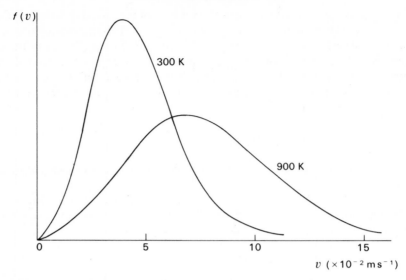

Fig. 8.2   The speed distribution function for a perfect gas (oxygen) at 300 K and 900 K.

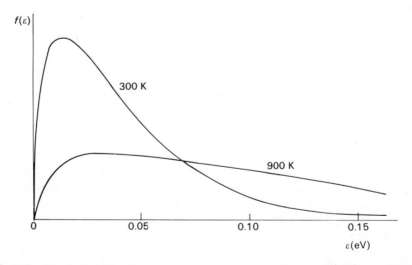

Fig. 8.3   The energy distribution function for a perfect gas at 300 K and 900 K. Notice that (unlike the other distribution functions) the energy distribution function is independent of the mass of a molecule of the gas and is a function only of temperature.

8.2 Mean values

The mean value of a quantity such as x^n is defined in terms of the distribution function to be

$$\overline{x^n} = \frac{\int x^n f(x) \, dx}{\int f(x) \, dx} \tag{8.9}$$

where as usual the integral extends over the whole range of x. The mean value of the velocity

$$\overline{v_x} = \frac{\displaystyle\int_{-\infty}^{\infty} v_x e^{-\frac{mv_x^2}{2kT}} \, dv_x}{\displaystyle\int_{-\infty}^{\infty} e^{-\frac{mv_x^2}{2kT}} \, dv_x} = 0 \tag{8.10}$$

as may be seen immediately since the numerator is the product of an odd function with an even function. Similarly the other components of the velocity are zero, and in general

$$\overline{v^n} = 0 \quad (n \text{ odd}). \tag{8.11}$$

The mean square value of the velocity is found from

$$\overline{v_x^2} = \frac{\displaystyle\int_{-\infty}^{\infty} v_x^2 e^{-\frac{mv_x^2}{2kT}} \, dv_x}{\displaystyle\int_{-\infty}^{\infty} e^{-\frac{mv_x^2}{2kT}} \, dv_x} = \frac{kT}{m} \tag{8.12}$$

$$\overline{v^2} = \overline{v_x^2} + \overline{v_y^2} + \overline{v_z^2} = \frac{3kT}{m}. \tag{8.13}$$

The mean square width of the velocity distribution may be found from the general result

$$\overline{(v - \bar{v})^2} = \overline{v^2} - (\bar{v})^2 = \overline{v^2} = \frac{3kT}{m}. \tag{8.14}$$

The width of the distribution for a given gas is therefore proportional to the square root of the absolute temperature.

The root mean square velocity of a molecule is of the same order as the speed of sound in the gas ($\approx 500 \text{ m s}^{-1}$ at room temperature) since

$$c_s = \sqrt{\frac{\gamma P}{\rho}} = \sqrt{\frac{\gamma kT}{m}} \tag{8.15}$$

where γ is the ratio C_P/C_V equal to 5/3 for a monatomic gas.

The mean speed is found using eqn 8.7

$$\bar{v} = \frac{\int_0^\infty v^3 e^{-\frac{mv^2}{2kT}} dv}{\int_0^\infty v^2 e^{-\frac{mv^2}{2kT}} dv} = \sqrt{\frac{8kT}{\pi m}}. \tag{8.16}$$

The integral now runs only from zero to infinity since speed is a scalar quantity. The mean square speed is

$$\overline{v^2} = \frac{\int_0^\infty v^4 e^{-\frac{mv^2}{2kT}} dv}{\int_0^\infty v^2 e^{-\frac{mv^2}{2kT}} dv} = \frac{3kT}{m} \tag{8.17}$$

in agreement with eqn 8.13.

The mean translational energy of a particle may be found from eqn 8.8 or from the equation

$$\bar{\varepsilon} = \tfrac{1}{2} m \overline{v^2} = \tfrac{3}{2} kT \tag{8.18}$$

a result that was obtained previously as an example of the equipartition theorem.

The important conclusion from these results is that the value found for any physical quantity (the coefficient of viscosity for example) will depend upon the way in which average values are taken. An incorrect average may lead to a relatively small numerical error, the speed values found so far for example are in the ratio

$$\sqrt{\overline{v^2}} : \bar{v} : v_{mp} = 1.22 : 1.13 : 1. \tag{8.19}$$

On the other hand, it may lead to complete nonsense as when the mean and mean square velocities are confused. An average value must be taken as late as possible in the calculation as will become apparent from the calculation in Section 8.4 of the pressure exerted by a perfect gas.

8.3 Doppler broadening of spectral lines

The radiation emitted by the molecules in a discharge tube is observed to have a finite width due to the random motion of the molecules. The motion of the molecules along the line of sight leads to Doppler broadening of the spectral lines. The transverse motion does not lead to Doppler broadening provided (as is always the case in the laboratory) that the velocity of the molecules is much less than the speed of light. The width of the spectral line may therefore be found from the one-dimensional velocity distribution function.

The intensity of the radiation emitted by molecules with velocity v_x to $v_x + dv_x$ may be written

$$I(v_x)\,dv_x = \alpha N f(v_x)\,dv_x \qquad (8.20)$$

where α is the radiation emitted per molecule. The wavelength of the light emitted by a molecule moving away from the observer with velocity v_x is given by

$$\lambda = \lambda_0\left(1 + \frac{v_x}{c}\right)$$

where λ_0 is the wavelength of the light emitted by a molecule at rest. The intensity of the radiation given by eqn 8.20 can therefore be related to the distribution as a function of wavelength since

$$v_x = c\left(\frac{\lambda - \lambda_0}{\lambda_0}\right); \quad dv_x = c\frac{d\lambda}{\lambda_0}$$

$$I(\lambda)\,d\lambda = \alpha N f(v_x)\left(\frac{dv_x}{d\lambda}\right)d\lambda$$

$$= \alpha N\left(\frac{m}{2\pi kT}\right)^{\frac{1}{2}}\exp\left[-\frac{mc^2}{2kT}\left(\frac{\lambda - \lambda_0}{\lambda_0}\right)^2\right]\frac{c}{\lambda_0}\,d\lambda.$$

The intensity is a maximum at the unshifted wavelength (λ_0) and has a Gaussian shape. The line may be normalized to the intensity at the centre of the line

$$I(\lambda) = I(\lambda_0)\exp-\frac{mc^2}{2kT}\left(\frac{\lambda - \lambda_0}{\lambda_0}\right)^2. \qquad (8.21)$$

The wavelengths for half the maximum intensity are given by

$$\ln 2 = \frac{mc^2}{2kT}\left(\frac{\lambda - \lambda_0}{\lambda_0}\right)^2$$

or

$$\lambda = \lambda_0 \pm \frac{\lambda_0}{c}\sqrt{\frac{2kT\ln 2}{m}}.$$

The full width of the line at half intensity is therefore

$$\Delta\lambda = \frac{2\lambda_0}{c}\sqrt{\frac{2kT\ln 2}{m}}. \qquad (8.22)$$

Since the mass of a molecule is fixed for a given gas the only way to reduce the Doppler broadening of a spectral line is to lower the temperature of the gas. An alternative approach is to allow the molecules to form a beam and to observe the radiation from the beam normal to the line of flight. The Doppler broadening is then unimportant but the intensity of the radiation is much lower than in a bulk sample of the gas.

Equation 8.22 has been used to measure the thermodynamic temperature of a plasma, i.e. an ionized gas. The electron temperature in a laser-heated plasma may be over 10^7 K and the measurement of such temperatures is a vital part of the work on the production of energy by nuclear fusion.

8.4 The passage of molecules across a plane surface

In this section an important general expression for the number of molecules crossing a plane area in the gas will be established and used to derive the velocity distribution function in an atomic beam and the perfect gas law.

The flow of molecules from one side through a plane surface of area S in the bulk of a gas in thermal equilibrium must be exactly balanced by the flow from the other side since there can be no preferred direction in the gas. The net flow is therefore zero and in the following discussion it is the flow of particles from one side only which is under consideration.

One special group of molecules will first be considered and the total number of molecules crossing the plane will be obtained at the end of the calculations. The number of molecules per unit volume of the gas with speed in the range v to $v+dv$ is (from eqn 8.7)

$$4\pi n \left(\frac{m}{2\pi kT}\right)^{\frac{3}{2}} v^2 \, e^{-\frac{mv^2}{2kT}} \, dv$$

where n is the total number of molecules per unit volume.

Since the gas is homogeneous, the fraction of the molecules moving in a direction defined by an infinitesimal solid angle $d\omega$ is simply $d\omega/4\pi$. Consider the situation shown in Fig. 8.4 where the element of solid angle makes an angle θ with the normal to the surface S. A molecule with speed v will cross

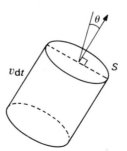

Fig. 8.4 A molecule with speed v travelling at an angle θ to the normal to the surface S will pass through the surface after a time dt if it was originally within the cylinder of base S and slant height vdt.

the surface S after a time dt if it is within a distance $v\,dt$ of the surface. The total number of molecules crossing the plane in time dt with speed v is equal to the number contained in a cylinder of base S and slant height $v\,dt$. The volume of the cylinder is therefore $vS\cos\theta\,dt$.

The number of molecules crossing S in time dt with speed in the range v to $v+dv$ in solid angle $d\omega$ making an angle θ with the normal to S is

$$4\pi n\left(\frac{m}{2\pi kT}\right)^{\frac{3}{2}} v^2\,e^{-\frac{mv^2}{2kT}}\,dv\left(\frac{d\omega}{4\pi}\right)(vS\cos\theta\,dt).$$

The element of solid angle may be written in polar coordinates

$$d\omega=\sin\theta\,d\theta\,d\phi$$

so the number of molecules crossing unit area in unit time may be written

$$dR=n\left(\frac{m}{2\pi kT}\right)^{\frac{3}{2}} v^3\,e^{-\frac{mv^2}{2kT}}\,dv\,\sin\theta\cos\theta\,d\theta\,d\phi. \qquad (8.23)$$

This is the general result from which the total number of molecules crossing the plane from one side and the pressure exerted by the gas may be evaluated. It is also the distribution of the molecules which emerge from a sufficiently small hole in the containing wall (an atomic beam, Section 9.2).

The total number of molecules crossing unit area in unit time from one side is found by integrating eqn 8.23 over all speeds and all allowed angles.

$$R=n\left(\frac{m}{2\pi kT}\right)^{\frac{3}{2}}\int_0^\infty v^3\,e^{-\frac{mv^2}{2kT}}\,dv\int_0^{\frac{\pi}{2}}\sin\theta\cos\theta\,d\theta\int_0^{2\pi} d\phi. \qquad (8.24)$$

The integration over θ is restricted to the range 0 to $\pi/2$ because molecules making an angle greater than $\pi/2$ with the normal to the plane are moving in the opposite direction from that under consideration. The net flux of molecules is found by allowing θ to run from 0 to π and is then seen to be correctly zero.

The first integral in eqn 8.24 is, from eqn 8.7, equal to the mean speed of the molecules (apart from a factor of 4π) and the angular integrals are equal to $\frac{1}{2}$ and 2π respectively.

The total number of molecules crossing unit area from one side in unit time is

$$R=\tfrac{1}{4}n\bar{v}. \qquad (8.25)$$

The average value $\bar{v}$ was correctly found at the end of the present calculation. The more elementary calculation which considers 1/6 of the molecules to be moving in one direction along any axis in the gas would obviously lead to a result $nv/6$ but the correct average for v is undetermined.

The importance of performing averages at the end of a calculation is also apparent from the calculation of the pressure exerted by a gas. In a dilute gas the number of molecules striking unit area of the wall of the container in unit time is identical to the number of molecules passing from one side through an imaginary surface in the interior of the gas. The momentum delivered normal to unit area of the wall in unit time by molecules with speed v moving at an angle θ to the normal to the wall is then, from eqn 8.23

$$mv \cos \theta \left[n \left(\frac{m}{2\pi kT} \right)^{\frac{3}{2}} v^3 e^{-\frac{mv^2}{2kT}} dv \sin \theta \cos \theta \, d\theta \, d\phi \right].$$

If the molecules remained attached to the wall after striking it the pressure exerted by the gas on the wall would simply be equal to

$$P = nm \left(\frac{m}{2\pi kT} \right)^{\frac{3}{2}} \int_0^\infty v^4 e^{-\frac{mv^2}{2kT}} dv \int_0^{\frac{\pi}{2}} \sin \theta \cos^2 \theta \, d\theta \int_0^{2\pi} d\phi \qquad (8.26)$$

$$= \frac{nm\overline{v^2}}{6} = \frac{nkT}{2} \qquad (8.27)$$

using the result of eqn 8.13 for $\overline{v^2}$.

When the molecules collide with the wall of the container and return to the bulk of the gas, the total pressure exerted must be twice that given by eqn 8.27 regardless of the nature of the collisions, so the usual result

$$P = \tfrac{1}{3} nm\overline{v^2} = nkT \qquad (8.28)$$

is obtained. A molecule could either make a specular reflection (angle of incidence equal to angle of reflection) with the wall or return to the gas at an angle independent of the angle of incidence. It was first shown by Knudsen that all (or nearly all) of the molecules return to the gas with a distribution identical to that with which they arrive (eqn 8.23). This point is considered in more detail in Section 9.4.3 but the pressure exerted by the molecules is independent of the nature of the surface provided that the molecules return to the bulk of the gas.

The importance of performing averages at the end of a calculation is apparent from eqn 8.26. An attempt to use the exact result for the total number of molecules passing through the plane (eqn 8.25) and some kind of average value for the momentum would not have been successful. (The roughest calculation involving 1/6 of the molecules moving in a given direction which was seen to be incorrect for the total number of molecules crossing the plane does however lead to eqn 8.28 due to the coincidence that the integral over the angle θ in eqn 8.26 is equal to 1/3.)

The form of eqn 8.25 is extremely important for surface physics. The clean surface of the material to be studied is normally prepared in a vacuum chamber with residual pressure P. The clean surface will be contaminated by molecules coming from the vacuum space at a rate given by eqn 8.25. Assuming that all the molecules remain on the surface and have diameter d

the time τ to completely cover the surface is given by

$$\left(\frac{1}{4}n\bar{v}\right)\left(\frac{\pi d^2}{4}\right)\tau \approx 1$$

τ may be related to the pressure using the gas law in the form $P = nkT$ and eqn 8.16 may be substituted for $\bar{v}$ to give

$$\tau \approx \frac{16}{\pi d^2}\sqrt{\frac{\pi m}{8kT}\frac{kT}{P}}. \tag{8.29}$$

This time is roughly 30 s (Exercise 8.3) for a pressure of 10^{-7} torr at room temperature (1 torr = pressure of 1 mm mercury = 133 N m^{-2}). Since a pressure of 10^{-7} torr is quite a good laboratory vacuum (an ultra-high vacuum is usually considered to be better than 10^{-8} torr) it is clear that surface contamination can be a serious problem in surface physics and has in the past led to many spurious results.

A further use of eqn 8.25 is in the measurement of small vapour pressures. The number of molecules of vapour striking the surface would be given by eqn 8.25 and this must also be the number of molecules of the solid which leave the surface to preserve the equilibrium. If all the molecules leaving the surface are collected then n can be determined and hence the vapour pressure (using eqn 8.28).

8.5 Effusion

When two gas containers are joined together by a pipe of large diameter the gas comes to an equilibrium condition in which the pressure is constant throughout the system. The gas flow between the containers while the system comes to equilibrium is essentially a hydrodynamic problem. If the temperature of one of the containers is now changed a further mass flow will occur until the pressures are again equalized. Two containers joined by a pipe of small diameter or a porous membrane however behave quite differently (Fig. 8.5).

When the aperture joining the two containers is sufficiently small the molecules in the region of the aperture may make a collision and pass out of the container without affecting the equilibrium of the rest of the gas. A small aperture is therefore one which has a dimension much less than the average distance a molecule travels between collisions (the mean free path, Section 9.1). The number of molecules passing through the aperture in this case will be given by eqn 8.25 provided that the thickness of the aperture is sufficiently small for the integrations in eqn 8.24 to be applicable.

The number of molecules passing from one container to another will, in the steady state, be given by

$$\tfrac{1}{4}n_A\bar{v}_A = \tfrac{1}{4}n_B\bar{v}_B.$$

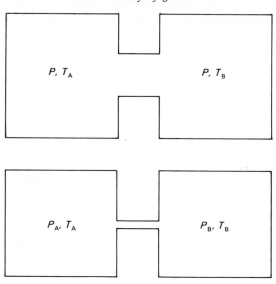

Fig. 8.5 Two containers joined by a pipe of large diameter always maintain the same pressure. Two containers joined by a pipe of small diameter or by a porous plug are only at the same pressure if their temperatures are equal.

Therefore

$$\frac{P_A \bar{v}_A}{T_A} = \frac{P_B \bar{v}_B}{T_B}.$$

The pressure in the system is therefore constant when both containers are at the same temperature but in general (substituting for $\bar{v}$ using eqn 8.16)

$$\frac{P_A}{\sqrt{T_A}} = \frac{P_B}{\sqrt{T_B}}. \tag{8.30}$$

The process of gas transport through small apertures is called effusion and the pressure difference produced by a temperature gradient is called a thermomolecular pressure difference. The importance of eqn 8.30 is that in low-temperature physics the temperature of the system is often measured in terms of the pressure of helium vapour. The mean free path in a gas at low pressure (Section 9.4.3) may become greater than the diameter of the coupling tube to the manometer and the measured pressure for a system at 1 K and a manometer at 300 K for example would then be in error by a factor of $\sqrt{300}$. The possibility of a thermomolecular pressure difference must therefore always be considered in systems which contain temperature gradients and low-pressure regions.

Knudsen first demonstrated that the thermomolecular pressure difference could be used to produce an absolute manometer for the measurement of pressures in the range $10^{-2} - 10^{-7}$ torr. The manometer is shown schematically in Fig. 8.6. The vane A is suspended from a quartz fibre so that the separation between A and the block B is much less than the mean free path in the gas. When the temperature of the block B is raised by a heater from temperature T to temperature T_1 a thermomolecular pressure difference will be set up and the vane will rotate. If the pressure in the system at temperature T is P and the pressure between the block and the vane is P_1

$$\frac{P_1}{P} = \sqrt{\frac{T_1 + T}{2T}} \tag{8.31}$$

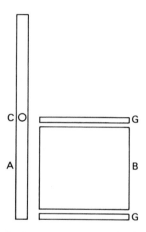

Fig. 8.6 Plan of Knudsen absolute manometer. The block B (surrounded by guard ring G) is heated causing the vane A (suspended from a fibre C) to rotate.

where the gas temperature in the gap is assumed to be the average of T and T_1.

When $(T_1 - T)$ is much less than T the equation may be simplified using the binomial theorem

$$\frac{P_1}{P} = \left(1 + \frac{T_1 - T}{2T}\right)^{\frac{1}{2}} = 1 + \frac{T_1 - T}{4T}.$$

The net force per unit area on the vane A is

$$P_1 - P = \left(\frac{T_1 - T}{4T}\right) P. \tag{8.32}$$

The force constant of the quartz fibre may be found in a separate experiment and then only the temperature of the system and the block B is required to

find the pressure in the system. An attractive feature of the Knudsen manometer is that the measured pressure is independent of the gas in the system.
The manometer is said to be 'absolute' because it measures pressure directly
in terms of the mechanical properties of the quartz fibre (which can be
determined in a separate experiment) rather than by comparing two pressures
as is commonly the case with, for example, the Pirani gauge (Section 9.4.3).
The lower limit to the useful pressure range of the Knudsen manometer is
about 10^{-7} torr. At lower pressures the torque due to the thermomolecular
pressure difference becomes so small that it is difficult to measure. At the
upper limit (10^{-2} torr) the separation between the plates cannot be made
much less than the mean free path in the gas and eqn 8.30 no longer applies.

One further important use of eqn 8.25 will be discussed in this section.
When two different gases are enclosed in a porous container the initial
effusion rates will depend upon both the concentration and the mass of the
molecules. If the molecules are of type A and type B and there are initially n_A
molecules of A and n_B molecules of B per unit volume then the ratio of the
molecules in the container is n_A/n_B but after effusion the ratio outside the
container is $n_A \bar{v}_A/n_B \bar{v}_B$ or substituting for the mean speed from eqn 8.16

$$\frac{R_A}{R_B} = \frac{n_A}{n_B} \sqrt{\frac{m_B}{m_A}}. \tag{8.33}$$

The gas inside the container is now richer in the heavier component than it
was before the effusion process started. Only a partial separation of the two
gases is possible in a single effusion process but the enriched gas may be used
in a cascade process to obtain complete separation. This method was used for
the first separation of the isotopes of uranium with the uranium in the form of
the gas UF_6.

Exercises

8.1. Find expressions for $\overline{v^3}$ and $\bar{v}^3$.

8.2. The mass of an atom of a monatomic gas is 1.4×10^{-25} kg. Show that the
fractional width of the Doppler broadened spectral line is 2×10^{-6} when
the gas is at a temperature of 650 K.

8.3. A clean metal surface is placed in a vacuum which has an oxygen
pressure of 10^{-7} torr at room temperature. Estimate the time taken to
cover the surface with oxygen if the diameter of an oxygen molecule is
0.23 nm and the system is at room temperature. How long would it take to
half cover the surface if the pressure was reduced to 10^{-10} torr? Assume all
the molecules that reach the surface remain on it, a reasonable assumption
at room temperature.

8.4. Show that the mean square velocity of a molecule emerging from a small hole in a gas container at temperature T is equal to $4kT/m$. Find the mean energy of the molecule and comment on the difference between this and the mean energy of a molecule inside the container.

8.5. Find an equation for the ratio of the concentrations of the two molecular species in a porous container during an effusion process if the gas that has effused is pumped away from the container. Hence show that no complete separation of the two species is possible in a single operation.

8.6. Find an expression for the number of electrons emitted from unit area of the surface of a metal in unit time if all electrons which strike the surface with velocity greater than v_0 normal to the surface escape and the electrons may be treated as a perfect gas.

8.7. Calculate the value of: (a) the mean speed; (b) the most probable speed; (c) the most probable velocity; (d) the r.m.s. velocity; and (e) the mean energy of oxygen and hydrogen molecules at 300 K. Calculate the velocity of sound in each gas.

8.8. At what temperature is the r.m.s. velocity of hydrogen molecules equal to the escape velocity from the earth? Explain why there is essentially no hydrogen in the atmosphere, (radius of the earth 6.4×10^6 m, acceleration due to gravity 9.8 m s^{-2}).

8.9. A plasma consists of a mixture of neutral atoms, positive ions and free electrons. The energy required to ionize an atom is typically a few electron volts. Estimate the temperature to which a gas must be heated to form a plasma.

8.10. Find an expression for the r.m.s. fluctuation of: (a) the velocity; (b) the speed; and (c) the energy of a single molecule in a monatomic gas. How are these expressions modified for the whole gas of N molecules?

8.11. What is the most probable kinetic energy of an atom in a gas at temperature T? Does it depend upon the mass of the atom? Is it equal to $mv_{mp}^2/2$?

8.12. Calculate the number of molecules of nitrogen which pass through 1 cm^2 of surface in 1 s at a temperature of 300 K when the pressure is (a) 1 atmosphere, (b) 10^{-6} N m^{-2}.

8.13. Show that $3PV = 2\bar{E}$ for a gas, provided that the energy of a molecule is given by $\varepsilon = p^2/2\,m$.

8.14. Find an expression for (a) the energy distribution function of the molecules of a gas in a molecular beam and (b) the fraction of the molecules which strike a surface with energy $\varepsilon > \varepsilon_0$. A molecule dissociates if it strikes a surface with energy greater than 0.6 eV. Show that the fraction of the molecules which dissociate doubles when the temperature increases from 300 to 310 K.

8.15. Find the value of the ratio given in eqn 8.33 for: (a) hydrogen and deuterium and (b) U^{238} and U^{235}, assuming that the initial number of molecules of each type of molecule is the same. What is the relative difficulty of separating the mixtures in (a) and (b)?

8.16. The pressure of the vapour of a substance in equilibrium with its liquid is called the saturated vapour pressure (SVP). The SVP for a material such as mercury may be found by pumping the vapour away continuously through a small aperture of area A and finding the change in the mass of the liquid after time t. Find an expression for the change in mass from which the SVP may be calculated. After 30 days at 273 K it is found that the mass of a drop of mercury (atomic weight 201×10^{-3} kg mol^{-1}) has decreased by 24×10^{-6} kg. Find the vapour pressure of mercury at 273 K if $A = 10^{-7}$ m^2.

9

Kinetic theory of gases—II

9.1 The mean free path

The concept of an average distance that a molecule travels between collisions (a mean free path) is not required in the rigorous transport theory of the classical perfect gas but remains a useful concept at the more elementary level adopted in this chapter. The mean free path is well defined for a gas of elastic spheres but requires more careful consideration when the more realistic model of a molecule as a scattering potential is used. Some values of the mean free path in real gases are given in Appendix VII.

Consider first an elastic sphere of radius b travelling with speed v through a random arrangement of stationary spheres of radius a. When the moving sphere makes a collision with one of the stationary spheres the centres of the two spheres are at a distance $(a+b)$ (Fig. 9.1). The direction of the motion of the moving sphere will change after each collision but in a time t (long compared to the time between collisions) the total distance travelled is simply vt. The volume swept out by the moving sphere in time t is $\pi vt(a+b)^2$. If the average number of spheres of type A per unit volume is n

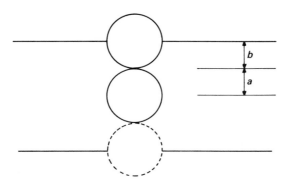

Fig. 9.1 A sphere of radius b sweeps out a volume with cross section equal to $\pi(a+b)^2$ in a distribution of spheres of radius a.

$$\lambda_1 = \text{mean free path} = \frac{\text{distance travelled in time } t}{\text{number of collisions in time } t}$$

$$= \frac{vt}{n\pi(a+b)^2 vt} = \frac{1}{n\pi d_{av}^2} \tag{9.1}$$

where

$$d_{av} = \tfrac{1}{2}(2a + 2b)$$

is the average diameter of the two types of sphere. The quantity

$$\sigma = \pi d_{av}^2 \tag{9.2}$$

is called the scattering cross-section.

The mean free path of a particle in a real gas of moving molecules is still given by eqn 9.1 if the particle of interest is moving much faster than the mean speed of the gas molecules. An electron in thermal equilibrium at temperature T will be moving much faster than a molecule of mass m since from eqn 8.13

$$\tfrac{1}{2}m_e \overline{v_e^2} = \tfrac{1}{2}m\overline{v^2} = \tfrac{3}{2}kT$$

and $m \gg m_e$. The diameter of an electron is also found to be negligible relative to that of a molecule so eqn 9.1 becomes

$$\lambda_e = (n\pi a^2)^{-1}. \tag{9.3}$$

The calculation of the mean of all the free paths in a gas of elastic spheres moving with a Maxwellian distribution of velocities is straightforward but the proof is tedious and will be omitted here. It is possible in fact to define more than one type of average in this case but the most commonly used (sometimes called the Maxwell free path) is

$$\lambda = \frac{\text{total distance travelled by the molecules in time } t}{2 \times \text{number of collisions in time } t} \tag{9.4}$$

$$= \frac{\lambda_1}{\sqrt{2}}. \tag{9.5}$$

The factor of 2 in eqn 9.4 is required because each collision terminates two free paths in the gas.

The mean free path of a molecule in a gas is clearly a function of its speed. A stationary molecule has a mean free path of zero, a very fast molecule a mean free path given by λ_1 and a typical molecule a mean free path of λ. The Maxwell mean free path may be written in a number of equivalent forms using the gas laws. In a gas containing only one type of molecule of diameter d, for example

$$\lambda = \frac{1}{\sqrt{2}\,n\pi d^2} = \frac{kT}{\sqrt{2}\,\pi d^2 p}. \tag{9.6}$$

The mean free path of a molecule in a given gas therefore varies such that

$$\lambda \propto n^{-1} \propto \text{density}^{-1}. \tag{9.7}$$

The diameter of a gas molecule (found from the measurement of the coefficient of viscosity (Section 9.4.1) is typically 0.3 nm so at NTP the Maxwell mean free path is about 10^2 nm. The mean free path is much greater than a molecular diameter but is also much less than the dimensions of a normal container. The molecules therefore spend most of their time in free flight but normally end a free path by colliding with another molecule rather than the wall of the container. When the mean free path becomes comparable with the dimensions of the system the interaction of the gas molecules with the surface of the container has to be considered (Section 9.4.3). The molecules collide on average after a time roughly equal to that required to travel a mean free path at the average speed. At NTP this time is about 10^{-9} s. In Chapter 1 it was seen that from computer studies it appears that a gas with a non-equilibrium distribution of velocities approaches equilibrium after two collisions per molecule. An equilibrium velocity distribution is therefore established at NTP in a time short compared to most experimental measuring times but at low pressure and temperature the relaxation time will increase.

At first sight it appears to be quite unrealistic to attempt to represent the molecules of a gas by a set of hard elastic spheres. The theory of the scattering of one atomic particle by another forms an important branch of quantum theory and the general scattering problem can certainly not be reduced to the single energy-independent cross-section found for elastic spheres. The subtleties associated with the detailed quantum theory of atomic scattering however turn out to be of most importance when the incident particle is only slightly deflected from its original path (low-angle scattering). The mean free path of electrons which are scattered out of a narrow beam is therefore energy dependent. In transport theory however the molecules which make the most important contribution to the coefficient of viscosity or the thermal conductivity are found to be those which are scattered through large angles and an energy-independent cross-section is found to be a much better approximation for large-angle scattering than for small-angle scattering. The cross-sections found from atomic beam measurments would therefore be expected to be greater than those found from measurements of the transport coefficients.

The effective cross-sections for gas molecules, calculated using the measured transport coefficients and the kinetic theory discussed in Section 9.4.1 are found to be temperature dependent showing that the elastic hard sphere model cannot be exact but the dimensions found for the molecules are of the right order of magnitude.

9.2 Atomic beams

The atoms of a gas may be allowed to escape from an oven at temperature T through an aperture which is sufficiently small not to disturb the equilibrium in the bulk of this gas (Section 8.5). If the space outside the oven is evacuated to a pressure of about 10^{-7} torr the molecules will travel a mean free path of about 1 m before colliding with a molecule of the residual gas in the vacuum. The velocity distribution in the beam will be given by eqn 8.23 provided that the thickness of the defining slits of the aperture in the oven is sufficiently small. The temperature to be used in eqn 8.23 is that of the oven from which the beam emerged. The beam cannot be said to have a temperature since it is not in thermal equilibrium but has a 'memory' of the oven temperature until the molecules collide with the residual gas molecules in the vacuum. An atomic beam apparatus may be used to verify the Maxwell velocity distribution in the form given by eqn 8.23 (Section 9.3) but in a normal apparatus the beam emerges from the oven not through a thin slit but from a long pipe. This increases the flux of molecules in the forward direction but the pressure in the oven must be lowered until the mean free path in the gas is of the order of the length of the pipe.

The attenuation of an atomic beam may be expressed in terms of the mean free path in the vacuum space. The simplest case to consider is that of a beam of particles of speed v each of which is thrown out of the beam after making one collision with a molecule of the residual gas in the vacuum space. If the number of particles left in the beam after traversing a distance x in the vacuum is $N(x)$ the number scattered in a length $\mathrm{d}x$ will be proportional to $N(x)$ and may be written

$$\mathrm{d}N = -\alpha N(x)\,\mathrm{d}x \tag{9.8}$$

where α is independent of x and is made into a positive quantity by the negative sign on the right-hand side of the equation. The solution of eqn 9.8 is

$$N(x) = N(0)\,\mathrm{e}^{-\alpha x} \tag{9.9}$$

where $N(0)$ is the initial number of molecules in the beam. The probability of a particle travelling for a distance x without making a collision is

$$P(x) = \mathrm{e}^{-\alpha x}.$$

The mean free path of a particle in the beam is given by

$$\lambda_v = \frac{\displaystyle\int_0^\infty x\,\mathrm{e}^{-\alpha x}\,\mathrm{d}x}{\displaystyle\int_0^\infty \mathrm{e}^{-\alpha x}\,\mathrm{d}x} = \frac{1}{\alpha}$$

so the final form for the attenuation of the beam may be written

$$N(x) = N(0)\,\mathrm{e}^{-\frac{x}{\lambda_v}}. \tag{9.10}$$

The scattering cross-section may therefore be found by measuring the attenuation of a beam as a function of distance. If the particles in the beam are passed through a velocity selector before they reach the vacuum space, the cross-section may be studied as a function of energy. The early studies of electron-molecule scattering using this method first showed that the scattering cross-section was a function of the electron energy rather than a constant as suggested by eqn 9.1.

Apart from their obvious importance in pure physics, atomic beams are now used in applied physics and in chemistry. The most accurate frequency standard is based on the measurement of an atomic transition in a beam of Cs atoms and semiconductor devices (microchips) are now fabricated using molecular beam epitaxy (MBE) as discussed in Exercise 9.14. The reaction dynamics of pairs of molecules with defined energy has been studied using crossed molecular beams and lead to the award of the 1986 Nobel Prize in Chemistry to Herschbach, Lee and Polanyi.

9.3 The verification of the Maxwell velocity distribution

An atomic beam apparatus combined with a velocity selector [Fig. 9.2(a)] may be used to verify the Maxwell velocity distribution in the form appropriate for the molecules effusing from an oven (eqn 8.23)

$$f(v)\,\mathrm{d}v = Av^3\,\mathrm{e}^{-\frac{mv^2}{2kT}}\,\mathrm{d}v \tag{9.11}$$

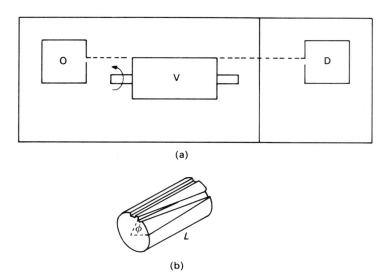

(a)

(b)

Fig. 9.2   (a) A schematic diagram of an atomic beam apparatus. The vapour leaves the oven (O) and passes into a vacuum space containing a velocity selector (V) and a detector (D). (b) Details of velocity selector.

where A is a constant for a given gas and temperature. The velocity selector could be of the notched-wheel type used in the classic experiments of Fizeau on the velocity of light but a more effective velocity selector is formed by cutting a series of slots in a cylinder, as shown in the diagram. When the cylinder is rotating with angular velocity ω a molecule with speed v will travel along the slot without changing its position relative to the sides of the slot if

$$v = \frac{\omega L}{\phi_0}$$

where L is the length of the cylinder and ϕ_0 is defined in Fig. 9.2(b). Molecules with a range of speeds will pass through the slots at a given value of ω since the slots have finite width. It may be shown that the range of speeds is proportional to the central speed v_0 and may therefore be written γv_0 where γ is a constant. The intensity of the particles incident on the detector is therefore given by eqn 9.11 multiplied by the factor γv_0 or

$$I(v_0) = \text{constant } v_0{}^4 \, e^{-\frac{mv_0{}^2}{2kT}}. \tag{9.12}$$

The equation may be written in a universal form (independent of temperature) in terms of the velocity ratio V defined by

$$V = \frac{v_0}{\text{predicted velocity at maximum intensity}}$$

when the intensity becomes

$$I(V) = 8\gamma V^4 \, e^{-2V^2}. \tag{9.13}$$

The agreement between this equation and the experimental result for potassium vapour is illustrated in Fig. 9.3. The best agreement is found when the defining slits of the aperture in the oven are as thin as possible, and the overall agreement is obviously good but the experimental results indicate that fewer slow molecules passed through the system than would be expected from eqn 9.11 unless the vapour pressure in the oven is as low as possible. This result is understandable since very slow molecules spend a comparatively long time in the vicinity of the oven slits and may therefore be scattered out of the beam before reaching the velocity selector.

The agreement between the best experimental results (obtained for thin oven slits and a low vapour pressure in the oven) is extremely striking when it is considered that there are no adjustable parameters at all in eqn 9.13. The theoretical velocity at maximum intensity is given from eqn 9.12 by

$$v_{\text{max}} = 2\sqrt{\frac{kT}{m}}$$

and may be calculated from the oven temperature T. The quantity V is therefore defined and the quantity γ is determined from the geometry of the velocity selector.

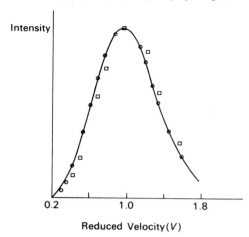

Fig. 9.3 The measured velocity distribution for potassium vapour. The Maxwell distribution is shown by the full line. The best agreement with theory was found in experiments using thin oven slits and low vapour pressures (shown by circles). (Miller and Kusch, 1955.)

The Maxwell velocity distribution may be considered to be well verified by both atomic beam experiments and also more indirectly by experiments such as the measurement of the Doppler broadening of spectral lines (Section 8.3).

9.4 Transport properties of a perfect gas

All the processes discussed so far in this book have related to systems which were either in true thermal equilibrium or (as in the case of atomic beams) could be related to a system which was effectively in thermal equilibrium. The basic concept of thermodynamics and equilibrium statistical mechanics has been seen to be that any isolated system will change from an arbitrary initial state to a final equilibrium state whose properties are independent of time. When gravitational effects are neglected a system in thermodynamic equilibrium has no preferred direction in space. The number of molecules crossing unit area of a surface in unit time was seen to be $n\bar{v}/4$ (eqn 8.25) if transport from only one side was considered, but the net flux of molecules was zero.

A more complicated situation arises when the system is not isolated but is in contact with reservoirs which drive the system from the equilibrium state. For example a pressure gradient along a tube will lead to a net flux of particles towards the low-pressure reservoir, a temperature gradient will lead to a net flux of energy and possibly also a net flux of particles and so on. An isolated system comes to thermal equilibrium in some characteristic time (the relaxation time) after which its properties are independent of time. Similarly it is found that the properties of a system which has a temperature gradient maintained across it also become time independent after some characteristic time. The system is then said to be in a steady state.

The steady-state properties of a perfect gas will be considered in the next two sections but it must be emphasised that the theory of the steady state is a vastly more difficult problem than that of the equilibrium state. The properties of any system in an equilibrium state can be found in principle using the canonical or grand canonical partition function. The methods used for the calculation of the transport properties of the perfect gas in this book cannot however be extended to systems of strongly interacting particles. In recent years it has been shown by more advanced methods that the transport properties of any system at given temperature may be expressed in terms of the fluctuations of the system about its equilibrium mean values. In this sense it is possible to write expressions for the transport properties (Kubo formulae) which are as general as the partition function results for equilibrium properties, but the actual evaluation of, say, the thermal conductivity of a metal has not been made any easier by the general result.

The transport properties of a gas depend upon the ratio of the mean free path in the gas (λ) to the smallest dimensions of the container (a) and the molecular diameter (d). Four separate regions may be identified.

Region I $a \gg \lambda \gtrsim d$ (dense gas)

The theory of transport processes in dense gases and in liquids is an active branch of current research but will not be considered in this book.

Region II $a \gg \lambda \gg d$ (normal transport properties)

A gas at NTP has been seen in Section 9.1 to satisfy these conditions. The overwhelming majority of the molecular collisions are between molecules. The collisions between the molecules and the surface of the container may safely be ignored when calculating the transport properties. The measured value of the transport coefficients in this region will be independent of the size of the container.

Region III $a \gtrsim \lambda \gg d$ (transition region)

The collisions of the molecules with the surface of the container are now of comparable importance to intermolecular collisions. The value of the coefficient of viscosity in this region [measured by the rate of flow of the gas along a tube (eqn 9.24)] now appears to depend upon the ratio a/λ. The theory of transport processes in this transition region will be briefly discussed after the results for region II have been obtained.

Region IV $\lambda > a \gg d$ (free molecular region).

The molecules now make more collisions with the container than with each other. The simplest approximate treatment is now to ignore intermolecular collisions completely. Region IV is of great importance for vacuum physics since the mean free path in a gas is ≈ 0.05 m at a pressure of 10^{-3} torr and is inversely proportional to the pressure at constant temperature. A good modern vacuum system operates at pressures in the range 10^{-8} to 10^{-14} torr where the conditions for region IV are obviously well satisfied. Similarly in outer space (where the effective pressure is only about 10^{-14} torr) the mean free path is much greater than the dimensions of a space vehicle.

9.4.1 Normal transport properties (region II)

The transport properties in this region are dominated by intermolecular collisions so collisions with the surface of the container will not be considered in this section. In an experiment to measure the thermal conductivity of a gas, a constant temperature gradient is maintained in the gas (Fig. 9.4). The experiment must be arranged so that no heat is transferred by mass flow (convection). The coefficient of thermal conductivity is defined by the equation

$$\dot{Q} = -\kappa A \frac{dT}{dx} \tag{9.14}$$

where $\dot{Q}$ is the energy transferred per unit time in the direction x normal to the surface of area A (Fig. 9.4).

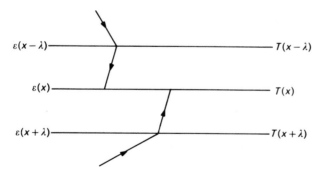

Fig. 9.4   In a thermal conductivity experiment a temperature gradient is maintained across the gas by external thermal reservoirs. The experiment must be arranged to prevent heat transport by convection. The molecules are assumed to make their last collision before crossing the reference plane at a distance equal to the mean free path in the gas and to take up the temperature (energy) of the molecules in this layer of gas.

In thermal equilibrium there is no nett flux of particles through the surface A. The basic concept of elementary transport theory is that at the microscopic level the temperature gradient (or any other driving mechanism) causes only a small perturbation of the system from the equilibrium state. The temperature gradient in an experiment might for example be $\approx 10^2$ K m^{-1}.

The mean free path in a gas at NTP is about 10^{-7} m. The 'temperature' change across a mean free path is therefore about 10^{-5} K which is very much less than the temperature of the gas. The transport coefficients defined by equations of the type 9.14 are therefore assumed to be independent of the magnitude of the driving gradient. It is possible in certain systems (very pure semiconductors at low temperature for example) to produce sufficiently large gradients to make this assumption incorrect but it may be safely employed for gases.

As has already been stressed the rigorous theory of transport in a perfect gas is mathematically difficult. An outline of the rigorous theory will be given in Section 9.5 but in the present section a very qualitative argument will be used to find expressions which are correct (apart from a numerical factor) for a gas of elastic spheres. The theory of polyatomic gases and gas mixtures will not be considered.

The elementary calculation of the quantity $\dot{Q}$ in eqn 9.14 may be made on the basis of two assumptions: (a) the number of molecules crossing unit area in unit time is $n\bar{v}/4$ as for a gas in thermal equilibrium and (b) a molecule which crosses the reference plane at x_0 makes on average a last collision a distance equal to the mean free path (λ) from the reference plane and takes up the temperature characteristic of the planes $(x+\lambda)$ or $(x-\lambda)$ (Fig. 9.4). The first assumption should immediately make the calculation suspect, since it involves using the integrated quantity $n\bar{v}/4$ rather than beginning by considering a group of molecules with speed in the range v to $v+dv$.

The energy transport across the area A in unit time may now be written

$$\dot{Q} = \tfrac{1}{4}\, n\bar{v}\, A[\bar{\varepsilon}(x+\lambda) - \bar{\varepsilon}(x-\lambda)]$$

where $\bar{\varepsilon}(x)$ is the mean energy of a molecule at x. Since the thermal gradient over a mean free path has been seen to be very small, the energy terms may be expanded and only the first two terms in the series retained.

$$\dot{Q} = \tfrac{1}{4}\, n\bar{v}\, A\left[\bar{\varepsilon}(x) + \lambda\frac{\mathrm{d}\bar{\varepsilon}}{\mathrm{d}x} - \bar{\varepsilon}(x) + \lambda\frac{\mathrm{d}\bar{\varepsilon}}{\mathrm{d}\lambda}\right]$$

$$= \tfrac{1}{2}\, n\bar{v}\, A\lambda\frac{\mathrm{d}\bar{\varepsilon}}{\mathrm{d}x}.$$

The mean energy of a molecule of a monatomic gas is

$$\bar{\varepsilon} = \tfrac{3}{2}kT$$

so

$$\frac{\mathrm{d}\bar{\varepsilon}}{\mathrm{d}x} = \frac{\mathrm{d}\bar{\varepsilon}}{\mathrm{d}T}\frac{\mathrm{d}T}{\mathrm{d}x} = c_V\frac{\mathrm{d}T}{\mathrm{d}x}$$

$$\dot{Q} = \tfrac{1}{2}\, n\bar{v}\, A\lambda c_V\frac{\mathrm{d}T}{\mathrm{d}x}$$

and by comparison with eqn 9.14

$$\kappa = \tfrac{1}{2} n \bar{v} \lambda \, c_V \qquad (9.15)$$

where c_V is the specific heat per molecule at constant volume. The factor of $\tfrac{1}{2}$ cannot of course be taken seriously after such a careless method of taking average values. A rather more careful calculation leads to the standard result (frequently used in solid-state physics)

$$\kappa = \tfrac{1}{3} n \bar{v} \lambda \, c_V \qquad (9.16)$$

but this too is incorrect. The rigorous result for a gas of hard spheres is

$$\kappa = 1.23 \, n \bar{v} \lambda \, c_V. \qquad (9.17)$$

All three results however have the same functional form. At constant temperature $\bar{v}$ is constant for a given gas and the product $n\lambda$ is always a constant for a gas of hard spheres (eqn 9.7) so κ is independent of n (gas density) at a given temperature. This surprising result (which will also be seen to be true for the coefficient of viscosity) is in excellent agreement with experiment for gases in region II and was one of the first successes of the kinetic theory of gases.

At constant density the thermal conductivity (according to eqn 9.17) is proportional to $\bar{v}$ and therefore to the square root of the temperature. This prediction agrees less well with experiment, a law of the form

$$\kappa \propto T^n \quad \text{(constant density)} \qquad (9.18)$$

with n about 0.7 being observed for monatomic gases. This result is understandable if the effective diameter of a molecule decreases with temperature. The molecules move faster at higher temperature and can therefore come closer together when making a collision. The 0.7-power law can be derived for various models of a molecule as a field of force. The simplest model involves only a repulsive force between the molecules of the form $1/r^n$ where n is about 11 for hydrogen and 15 for helium. The elastic sphere model would give a value of n equal to infinity. Many more elaborate laws of force have been used to improve the agreement between theory and experiment.

These calculations were of great importance in the early days of kinetic theory since they provided the first information about molecular sizes using, say, eqn 9.17 for the coefficient of thermal conductivity and eqn 9.6 to relate the mean free path to the molecular diameter. Molecular scattering now more naturally forms a part of quantum mechanics and will not be discussed further in this book.

A simple calculation will now be given for the coefficient of viscosity of a gas. The definition of the coefficient of viscosity may be given in terms of Fig. 9.5. The layer of gas next to the wall is assumed to be at rest and therefore exerts a force tangential to the surface on the next layer of gas which is moving. Assuming that the velocity of mass flow of the gas (U_x) is much less

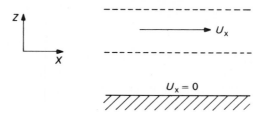

Fig. 9.5 When the velocity of mass flow in a gas (U_x) depends upon the distance (z) from a reference plane normal to the flow a force is exerted on each layer of gas by its neighbouring layers. The force per unit area is proportional to the gradient ($\partial U_x/\partial z$) if U_x is much less than the mean thermal speed of the molecules.

than the mean thermal speed of the gas molecules, the tangential force between the layers will be proportional to the velocity gradient. The force per unit area may therefore be written

$$F = -\eta \frac{\partial U_x}{\partial z} \tag{9.19}$$

where η is a constant called the coefficient of viscosity.

The calculation of the coefficient of viscosity is now carried out as for the thermal conductivity but it is now the net flux of momentum across the plane which is of interest, since by Newton's third law ($F = \mathrm{d}p/\mathrm{d}t$) this is equal to the net force exerted. Therefore

$$F = \tfrac{1}{4} n\bar{v} m \left[U_x(z+\lambda) - U_x(z-\lambda) \right]$$
$$= \tfrac{1}{4} n\bar{v} m \left[U_x(z) + \lambda \frac{\partial U_x}{\partial z} - U_x(z) + \lambda \frac{\partial U_x}{\partial z} \right].$$

Therefore

$$\eta = \tfrac{1}{2} n m \bar{v} \lambda = \tfrac{1}{2} \rho \bar{v} \lambda \tag{9.20}$$

where m is the mass of one molecule and ρ the density of the gas. A slightly better calculation again leads to a value of 1/3 rather than 1/2 for the numerical factor. The rigorous result for hard spheres is

$$\eta = 0.499 \, n m \bar{v} \lambda. \tag{9.21}$$

The coefficient of viscosity of a perfect gas in region II is therefore predicted to be independent of the density of the gas at constant temperature. This result was apparently first obtained experimentally by Boyle in 1660 (using the damping of the oscillations of a pendulum as a measure of the coefficient of viscosity) but was forgotten until Maxwell derived it theoretically and then confirmed it by experiment. The coefficient of viscosity is predicted by eqn 9.21 to be proportional to $\bar{v}$ and therefore to the square root of the absolute temperature but is found from experiment to be more nearly proportional to $T^{0.7}$, as was the thermal conductivity.

The combination of the equations for the thermal conductivity and coefficient of viscosity of a gas eliminates the mean free path and leads to one of the most striking predictions of the kinetic theory

$$\kappa = \alpha \eta \frac{c_V}{m} = \alpha \eta \frac{C_V}{M} \tag{9.22}$$

The coefficient α is equal to unity in the simple theory given in this section and to 2.52 in the rigorous theory of a hard-sphere gas. The value of α found from experiment using eqn 9.22 is therefore a direct check on kinetic theory calculations. The experimental values of α for monatomic gases are in quite good agreement with the value calcuated by the rigorous theory: the value of α for helium for example is 2.44 and for argon is 2.47. Diatomic gases give lower values, for example 1.92 for nitrogen. The discrepancy is attributed to the different contributions to the coefficient of viscosity and the thermal conductivity made by molecular rotations.

The good agreement between theory and experiment for the monatomic gases is not of course evidence that molecules may be treated as elastic spheres but rather that the effective size of a molecule in a thermal conductivity experiment is similar to that in a viscosity experiment. The calculated value of α for a gas of molecules interacting via a repulsive potential of the form $1/r^n$ for example is 2.50 which is almost identical to the result for a hard-sphere gas.

9.4.2 Gas flow along a tube (Poiseuille flow)

The flow of gas along a pumping tube in a system evacuated from room pressure is at first a flow process in region II and is controlled by the coefficient of viscosity. At lower pressure the mean free path in the gas becomes greater than the diameter of the pumping tube (region IV) and the concepts of the next section are required.

The gas flow in region II will be assumed to be laminar flow in the steady state, or as it is sometimes called, Poiseuille flow. A fast vacuum pump may in fact produce non-laminar flow in the pumping tube but this aspect will not be considered here. The variation in the drift velocity of a gas across a tube of circular cross-section and radius a may be found by considering the equilibrium of an imaginary cylinder in the gas (Fig. 9.6) under conditions of laminar flow. The viscous force on the curved surface of the cylinder is balanced by the difference between the forces acting on the two plane faces so

$$-\pi r^2 \frac{dP}{dx} = -2\pi r \eta \frac{dU_x}{dr}.$$

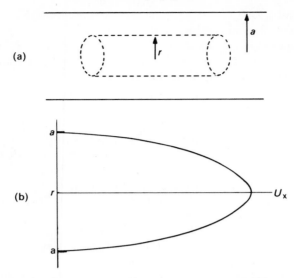

Fig. 9.6 (a) The equilibrium of an imaginary cylinder in the gas depends upon the forces exerted on the end faces by the pressure gradient being balanced by the frictional force due to viscosity exerted on the curved surface. Notice that the origin of the coordinates lies on the line through the centre of the cross-section of the tube. (b) The parabolic variation of the mass velocity across the tube.

Therefore

$$\int dU_x = \frac{1}{2\eta} \frac{dP}{dx} \int r \, dr$$

$$U_x = \frac{r^2}{4\eta} \frac{dP}{dx} + \text{constant.}$$

In region II it is assumed that the layer of gas next to the surface of the tube is at rest. Therefore U_x is zero when r is equal to a and

$$U_x = \frac{r^2 - a^2}{4\eta} \frac{dP}{dx}. \tag{9.23}$$

The variation of the drift velocity across the tube is therefore parabolic.

The total mass of gas flowing in unit time through a given cross-section in which the pressure is P is therefore

$$\dot{m} = \int_{r=0}^{a} \frac{P}{R_m T} 2\pi r \, U_x \, dr$$

using $P = \rho R_m T$ where R_m is the gas constant for unit mass of gas. The integration over r is straightforward after substituting for U_x and leads to a mass flow through a given cross-section of

$$\dot{m}_0 = \frac{-\pi a^4}{8\eta R_m T} P \frac{dP}{dx}.$$

The mass flow through each section of the tube must be a constant so after a further integration along the tube of length l

$$\dot{m}_0 = \frac{\pi a^4}{16\eta R_m T} \frac{P_1^2 - P_2^2}{l} = \frac{\pi a^4}{8\eta R_m T} \bar{P} \frac{\Delta P}{l} \tag{9.24}$$

where $\bar{P} = (P_1 + P_2)/2$ is the average pressure along the tube.

This equation shows the importance of connecting the system to be evacuated to a pump by tubing of short length and large diameter since a pump can at best produce P_2 equal to zero.

When the pressure gradient is small and the mean pressure is such that the gas in the tube may be considered to be in region II, a flow experiment and eqn 9.24 may be used to measure the coefficient of viscosity of a gas. As the mean pressure is lowered however the gas enters region III ($a \gtrsim \lambda \gg d$) and the coefficient of viscosity calculated using eqn 9.24 now becomes a function of the diameter of the tube. The equation is inadequate in region III because it is no longer permissible to assume that the gas in the layer next to the wall of the tube is at rest.

Knudsen suggested an improvement to eqn 9.24 based on the assumption that the mass velocity of the gas was such that it would extrapolate to zero at a distance equal to $(a + \lambda)$ from the centre of the tube where λ is the mean free path in the gas. In region II the mean free path is sufficiently small relative to the tube diameter for this correction to be negligible but this is not the case in region III. The mass flow per second may now be shown to be

$$\dot{m}_0 = \frac{\pi a^4}{8\eta R_m T} \left(\bar{P} + \frac{4\tau_1}{a} \right) \frac{\Delta P}{l} \tag{9.25}$$

where τ_1 is a constant at a given temperature equal to λP. The second term in the bracket obviously becomes of increasing importance for a given tube as the mean pressure in the tube decreases and at a given mean pressure it is more important for small tubes than large ones. It may be shown that the form of τ_1 implies that the surface of the tube appears completely rough to a molecule so that it is reflected in a direction which is independent of the direction from which it arrived on the surface. This point is discussed further in the next section.

As the mean pressure in the tube decreases into region IV ($\lambda > a \gg d$) eqn 9.25 also ceases to apply because the whole concept of layers of gas with different mass velocities is now invalid. The molecules now make more collisions with the walls of the tube than with each other. The mass flow for a given tube and pressure gradient is found to decrease with the mean pressure in agreement with eqn 9.25 but then passes through a minimum as shown in Fig. 9.7. At still lower pressure a new process leads to an increase in the mass flow. This free molecular flow in region IV will be discussed in the next section.

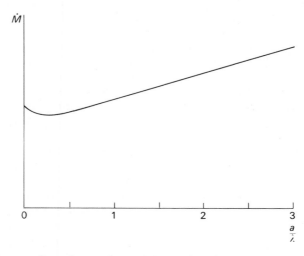

Fig. 9.7 The mass flow of a gas along a tube as a function of the mean pressure shows a minimum.

9.4.3 Transport properties at low density (region IV)

In region IV the gas molecules make more collisions with the walls of the container than intermolecular collisions. The simplest approximate treatment of this region therefore consists of ignoring intermolecular collisions completely and concentrating attention on the interaction of gas molecules with a surface.

A molecule which strikes a surface may return to the bulk of the gas immediately (after making an elastic collision) in a direction such that the angle of incidence of the molecule is equal to the angle of reflection. In optics this process is called specular reflection. The condition for specular reflection is that

$$\lambda > d \sin \theta \tag{9.26}$$

where λ is the wavelength of the radiation, θ the glancing angle with the surface and d the height of a typical surface irregularity. A gas molecule may be described by a de Broglie wave and it may easily be shown (Exercise 9.5) that even for a good polished surface ($d \approx 10^{-7}$ m) specular reflection is only possible for molecules with very small glancing angles. The best surfaces are of course the faces of crystals for which $d \approx 10^{-10}$ m and specular reflection may then be observed up to glancing angles of about $30°$ for light atoms.

The majority of the molecules leaving the surface of a tube would therefore be expected to return to the gas in a direction independent of the angle of incidence (diffuse reflection) because of the roughness of the surface, but diffuse reflection will also result from any surface if a molecule remains in contact with the surface for some period of time before it returns to the gas. The experiments of Kundsen suggest that the equation deduced for the flux of molecules on to the wall of a container (eqn 8.23) is also correct for the flux of molecules returning to the bulk of the gas from the surface.

A simple calculation of the mass flow of molecules at very low density along a long tube is possible if intermolecular collisions are neglected and the drift velocity along the tube is much less than the mean thermal speed of the gas molecules. The velocity gradient across the tube which was of vital importance in regions II and III may now be neglected.

The number of molecules crossing unit area in unit time will be assumed to be $n\bar{v}/4$ as for a gas in thermal equilibrium. The total momentum in the direction of gas flow given up to the walls of a tube of length l and radius a in unit time if the surface of the tube is rough is

$$(2\pi a l)\left(\frac{n\bar{v}}{4}\right)mU_0 \tag{9.27}$$

where U_0 is the drift velocity down the tube. The force given by eqn 9.27 must be balanced by the pressure difference across the ends of the tube so

$$2\pi a l \frac{n\bar{v}}{4} mU_0 = \pi a^2 (P_1 - P_2).$$

The drift velocity is therefore

$$U_0 = \frac{2a(P_1 - P_2)}{nm\bar{v}l}$$

and the total mass flow in unit time is found from the number of molecules in a cylinder of radius a and length numerically equal to U_0.

$$\dot{m}_0 = \pi a^2 nmU_0$$

where m is the mass of a molecule. Therefore

$$\dot{m}_0 = \frac{2\pi a^3}{\bar{v}} \frac{\Delta P}{l}. \tag{9.28}$$

The mass flow along a tube when the pressure is sufficiently low for the mean free path to be greater than the diameter of the tube is independent of the density of the gas and of the coefficient of viscosity.

This calculation for the mass flow along a long tube may be compared with the result for the effusion of a gas through a thin slit of radius a. The mass flow along the long tube is reduced by a factor of $(l/\pi a)$ relative to the effusion process, once again showing the importance of short pumping lines of large diameter.

The free molecular mass flow given by eqn 9.28 is much greater than that expected from Poiseuille flow at low pressure (Exercise 9.2). The pumping rate given by eqn 9.28 is a function of the mass of a molecule of the gas (through $\bar{v}$) so when a system is being evacuated the concentration of the residual gases will change with time (Exercise 9.3).

The problem of free molecular heat conduction may be treated in a similar way to free molecular mass flow but is complicated by the fact that in general the molecules returning to the bulk of the gas do not have a Maxwellian distribution of velocities corresponding to the temperature of the heated surface.

In a system in thermal equilibrium at temperature T the energy brought to unit area of surface in unit time by the molecules of a monatomic gas with speed between v and $v + dv$ is given by eqn 8.23 multiplied by $mv^2/2$. The total energy delivered to the surface is therefore

$$\frac{nm\overline{v^3}}{8}.$$

The value of $\overline{v^3}$ was found in Exercise 8.1 to be $4kT\bar{v}/m$ so the total energy delivered to unit area in unit time may be written

$$\dot{Q} = \tfrac{1}{2}n\bar{v}\,kT = \tfrac{1}{2}P\bar{v} \tag{9.29}$$

which for a system in thermal equilibrium must also be the energy transfer from the wall to the gas.

When a surface in a gas at temperature T is heated to temperature T_1 the net rate of transfer of heat per unit area is

$$\dot{Q} = \tfrac{1}{2}n\bar{v}k\,(T_1 - T) = P\bar{v}\left(\frac{T_1 - T}{2T}\right) \tag{9.30}$$

provided that the molecules actually leave the surface as if with temperature T_1. In general the molecules appear to leave with an effective temperature between T and T_1 and a correction must be made to the equation.

The Pirani gauge is a simple and useful pressure gauge based on eqn 9.30. A thin tungsten wire of diameter d and length l is heated electrically to temperature T_1. The temperature of the wire is found from its resistance (V/I) and the power dissipated in the wire is simply IV. Then

$$IV = \left[\frac{\pi dl(T_1 - T)}{2T}\sqrt{\frac{8kT}{\pi m}}\right]P \tag{9.31}$$

and the pressure may be calculated provided the gas in the system is known. The Pirani gauge is usually calibrated against a McLeod gauge rather than by using eqn 9.31 because the calculated pressure depends upon the type of molecules in the system and it is not known if the molecules leave the wire with velocities given by T_1. The Pirani gauge is useful in the pressure range 10^{-2} to 10^{-4} torr but at lower pressure the heat loss by radiation becomes more important than the free molecular heat conduction (Exercise 9.4).

9.5 The Boltzmann transport equation

The rigorous theory of the transport coefficients of a perfect gas was begun by Boltzmann and developed by Chapman and Enskog. A gas in region II is sufficiently dilute for only binary collisions of the molecules to be of importance and the transport theory may then be developed in terms of the single-particle distribution function discussed in earlier chapters. The theory of the dense gas however requires the distribution function for pairs of molecules and will not be considered here.

The generalized single-particle distribution function $f(v, r, t)$ is defined $f(v, r, t) \, d^3v \, d^3r$ is the average number of molecules with velocity in the range v to $v + dv$ and position in the range r to $r + dr$ at time t.

In an isolated system the generalised distribution function will tend to the equilibrium Maxwell distribution of Chapter 8 for all t greater than the slowest relaxation time in the gas.

The behaviour of the generalized distribution function in the absence of molecular collisions is governed by the equation (Liouville's equation of classical mechanics)

$$\frac{\partial f}{\partial t} + \frac{dr}{dt} \cdot \frac{\partial f}{\partial r} + \frac{dv}{dt} \cdot \frac{\partial f}{\partial v} = 0 \tag{9.32}$$

provided that the force acting on a particle is independent of its velocity. The force is then given by

$$F = m \frac{dv}{dt}$$

Equation 9.32 can be extended to include the effect of magnetic fields (which lead to velocity-dependent forces) but this will not be done here.

The equation may then be rewritten

$$\frac{\partial f}{\partial t} + v \cdot \frac{\partial f}{\partial r} + \frac{F}{m} \cdot \frac{\partial f}{\partial v} = 0 \tag{9.33}$$

in the absence of collisions.

The effect of interparticle collisions is to tend to make the system relax towards local thermal equilibrium. A collision term may be added to the right hand side of eqn 9.33 to give the Boltzmann transport equation in the form

$$\frac{\partial f}{\partial t} + \boldsymbol{v} \cdot \frac{\partial f}{\partial r} + \frac{\boldsymbol{F}}{m} \cdot \frac{\partial f}{\partial v} = \left(\frac{\partial f}{\partial t}\right)_{\text{coll}}.$$

The collision term was shown by Boltzmann to involve an integral over the (unknown) generalized distribution function. The Boltzmann transport equation is therefore an integro-differential equation which cannot be solved exactly.

The equation is normally simplified by assuming (as in Section 9.4) that the force acting on a particle is sufficiently weak for the transport coefficients to be independent of the force. The most general treatment of the Boltzmann equation (the variational method) which leads to expressions for the transport coefficients in terms of integrals which depend upon the law of force between the molecules, will not be considered here but instead a simpler treatment will be given.

The relaxation time approach replaces the collision term by the expression

$$\left(\frac{\partial f}{\partial t}\right)_{\text{coll}} = -\frac{f - f_0}{\tau} \tag{9.34}$$

where τ is the relaxation time and f_0 the equilibrium distribution. The effect of the collision term may be seen by 'turning off' the external force which has driven the system from equilibrium. Then

$$\frac{\partial f}{\partial t} = \frac{\partial (f - f_0)}{\partial t} = -\frac{f - f_0}{\tau}$$

since f_0 is independent of time, and the solution of the differential equation is

$$f(t) - f_0 = (f(0) - f_0) \, e^{-\frac{t}{\tau}}$$

where $f(0)$ is the value of the distribution function at the time when the force is switched off. The general distribution function therefore approaches the value for thermal equilibrium with characteristic time τ. The relaxation time approach can be rigorously justified for certain types of intermolecular interaction (Wannier (1966) Chapter 18) and is also often used in solid-state physics to give approximate results. A system will have a number of relaxation times (which may be functions of the energy of the particles) but the longest relaxation time is normally the one of interest.

The application of the Boltzmann transport equation in the form

$$\left(\frac{\partial f}{\partial t}\right) + \boldsymbol{v} \cdot \frac{\partial f}{\partial r} + \frac{\boldsymbol{F}}{m} \cdot \frac{\partial f}{\partial v} = \frac{f - f_0}{\tau} \tag{9.35}$$

with a single value for τ leads to values for the coefficient of viscosity and thermal conductivity with numerical coefficients of $1/3$ (as in eqn 9.16) which have already been seen to be inadequate. These calculations will not therefore be considered here but, as an illustration of the use of the transport equation, an expression for the electrical conductivity of a gas containing a low concentration of positive ions will be derived. This calculation has been chosen since it shows particularly clearly that the transport properties are due to the deviation of the system from thermal equilibrium.

The force term in eqn 9.35 may be written $eE_x x$ for a uniform electric field in the x-direction acting on a particle with charge e and the first term in the equation is zero if $E_x x$ does not change with time. The equation may now be written in the one-dimensional form

$$v_x \frac{\partial f}{\partial x} + \frac{eE_x}{m}\left(\frac{\partial f}{\partial v}\right) = -\frac{f-f_0}{\tau}$$

since a net current can only flow in the x-direction.

The generalized distribution function may therefore be written

$$f = f_0 - \tau\left(\frac{eE_x}{m}\frac{\partial f}{\partial v_x} + v_x\frac{\partial f}{\partial x}\right). \tag{9.36}$$

The distribution function before the application of the electric field was simply the Maxwell velocity distribution

$$f_0 = n\left(\frac{m}{2\pi kT}\right)^{\frac{3}{2}} e^{-\frac{mv^2}{2kT}} \tag{9.37}$$

where n is the number of ions per unit volume of the gas. The system is still at uniform temperature after the application of the field and if E_x is independent of position as well as time

$$\frac{\partial f}{\partial x} = 0.$$

The electrical conductivity (σ) is defined as the ratio of the current density (J_x) to the electric field

$$\sigma = \frac{J_x}{E_x} \tag{9.38}$$

where J_x (defined as the average electric charge crossing unit area in unit time) is given by

$$J_x = \int_{-\infty}^{\infty} ev_x f d^3v. \tag{9.39}$$

At small values of the electric field the electrical conductivity is independent of E_x (Ohm's law) and f must then be such that

$$|f - f_0| \ll f_0. \tag{9.40}$$

As a first approximation therefore the substitution

$$\frac{\partial f}{\partial v_x} \approx \frac{\partial f_0}{\partial v_x}$$

may be made on the right-hand of eqn 9.36. Equation 9.39 now becomes

$$J_x = \int_{-\infty}^{\infty} e v_x f_0 \, d^3v - \frac{e^2 E_x}{m} \int_{-\infty}^{\infty} v_x \tau \frac{\partial f_0}{\partial v_x} \, d^3v. \tag{9.41}$$

The first term on the right-hand side is equal to $n e \bar{v}_x$ and $\bar{v}_x$ is equal to zero (eqn 8.10). The current density is due to the second term, which represents the deviation of the system from equilibrium. The Maxwell distribution for f_0 (eqn 9.37) leads to

$$\frac{\partial f_0}{\partial v_x} = -\frac{m v_x}{kT} f_0$$

$$J_x = \frac{e^2 E_x}{kT} \int_{-\infty}^{\infty} \tau v_x^2 f_0 \, d^3v.$$

The integral cannot be evaluated without some assumption about the dependence of τ on the velocity (energy) of the particles but if it can be removed from the integral and represented by some average value $\bar{\tau}$ then

$$J_x = \frac{n e^2 E_x \bar{\tau}}{kT} \, \overline{v_x^2}$$

$$= \frac{n e^2 \bar{\tau}}{m} E_x$$

(using eqn 8.12) and

$$\sigma = \frac{J}{E_x} = \frac{n e^2 \bar{\tau}}{m}. \tag{9.42}$$

The electrical conductivity is independent of the field, in agreement with Ohm's law, provided that $\bar{\tau}$ is independent of the field.

The equation for the electrical conductivity could in fact have been obtained by the much simpler mean free path approach adopted in Section 9.4 (writing $\lambda \approx \bar{v} \bar{\tau}$ where $\bar{v}$ is the mean speed) but the derivation using the Boltzmann equation has clarified the condition (eqn 9.40) for which Ohm's law will apply. This inequality may be written (using eqn 9.36) in the form

$$kT \gg \tau v_x e E_x$$
$$kT \gg \lambda e E_x. \tag{9.43}$$

So Ohm's law is valid if the energy gained by a particle from the field between collisions is much less than its thermal energy. This condition was assumed throughout Section 9.4 but can now be seen to arise naturally in the solution to the Boltzmann equation.

The many assumptions and approximations that had to be introduced in the derivation of the electrical conductivity of a dilute gas demonstrates how much more difficult transport theory is than the study of equilibrium properties. The problem of transport theory in a dense gas or a liquid in which collisions of more than two particles cannot be ignored is obviously still more complicated and is an active branch of current research.

9.6 Conclusion

In Chapter 8 exact expressions were obtained for the thermal properties of a classical perfect gas in equilibrium. Only approximate values for the transport coefficients were obtained in this chapter but they are of more general interest than might be thought from the discussion to this point. The direct application of the kinetic theory to vacuum physics, surface physics and atomic beams is obviously of great importance, but in addition the perfect gas calculations have often been taken over into solid-state physics.

The application of the kinetic theory results for the electrical and thermal conductivities of a gas to the free electrons in a metal is discussed in Section 10.1. This is not a good example however because (as was seen in Section 6.5) the free electrons in a metal are not in the classical limit and must be treated by Fermi–Dirac statistics. The kinetic theory equation for the thermal conductivity of a gas (eqn 9.16)

$$\kappa = \tfrac{1}{3}nc_V\bar{v}\lambda$$

has however often been used to discuss the thermal conductivity of a solid due to lattice vibrations. A mean free path for the thermal conductivity is indeed often calculated from the measured thermal conductivity and specific heat but it should be clear from the discussion in Section 9.4 that the factor of 1/3 probably has little significance. The heat transport in an insulator may be thought of as carried by quasi-particles called phonons (in analogy with the particle-like (photon) aspect of the electromagnetic field) as is discussed in Sections 10.2 and 10.3 and the mean free path of a phonon increases at low temperature due to the decrease in the amplitude of lattice vibrations. The lattice thermal conductivity is therefore found to increase below room temperature. The mean free path cannot however exceed the diameter of the sample rod and therefore becomes constant at sufficiently low temperature. The kinetic theory result of eqn 9.16 is then sufficient to predict that the thermal conductivity of a pure insulator will become proportional to the product of the specific heat and sample diameter at sufficiently low temperature. Furthermore the lattice specific heat is proportional to T^3 at low temperature (Section 10.3) so the lattice thermal conductivity must show a maximum as a function of temperature (Fig. 9.8).

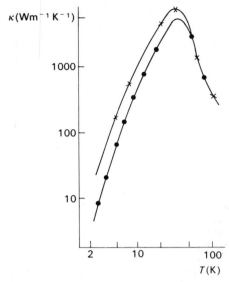

Fig. 9.8 The thermal conductivity of a pure crystal of an insulating solid (Al_2O_3) depends upon the diameter of the sample at low temperature: × 2.8 mm; ● 1.02 mm. After Berman, *et al.* 1955.

The results of the kinetic theory derived in Chapters 8 and 9 are therefore not only of use in calculating the properties of perfect gases but are also employed in other branches of physics.

Exercises

9.1. The coefficient of viscosity of helium at NTP is 1.86×10^{-5} kg m^{-1} s^{-1}. Find: (a) the thermal conductivity; (b) the mean free path; (c) the molecular diameter; and (d) the pressure at which the mean free path is equal to 10^{-2} m at 300 K and at 30 K.

9.2. Find the ratio of the mass flow along a tube expected for Poiseuille flow and free molecular flow when the mean free path is equal to the tube diameter.

9.3. Find an expression for the partial pressures of the residual gases in a vacuum system as a function of time in the region of free molecular flow. How does the time constant depend upon the mass of a molecule?

9.4. A Pirani gauge ceases to be useful when the heat loss from the wire due to radiation becomes comparable with the heat flow in the gas. Estimate this pressure if the wire radiates as a black body, the residual gas is oxygen, the system is at room temperature and the temperature difference between wire and gas is small.

9.5. Show that specular reflection is only possible at room temperature for hydrogen atoms with glancing angles less than 5 minutes of arc if the surface irregularities are of order 10^{-7} m.

9.6. Estimate the radius of an argon atom if liquid argon (density 1.4 g cm^{-3}) can be considered to consist of touching spherical atoms. Show that the mean spacing between the atoms in a gas at NTP is much greater than the radius of an atom.

9.7. Use the value of the radius of an argon atom found in the previous equation to estimate: (a) the mean free path of an electron; and (b) the mean free path of an argon atom in argon gas at NTP.

9.8. What is the attenuation factor for the scattering of a beam of argon by residual argon atoms at a pressure of 10^{-6} torr (133 μPa)?

9.9. In Section 9.4 we define four regions involving inequalities between the size of molecules, the mean free path and the size of the container. Estimate the pressures appropriate to each region for oxygen at room temperature assuming that the molecules can be considered as spheres of diameter 2.3 $\times 10^{-10}$ m and the container has a least dimension $\simeq 10^{-2}$ m.

9.10. One of the most important applications of kinetic theory is to high vacuum technology. The following questions relate to the problem of producing the vacuum and the contamination of surfaces in the vacuum by residual gases. An excellent reference for this topic is given in the bibliography (Weston).

(a) The gas throughout is defined as

$$Q = \frac{\mathrm{d}(PV)}{\mathrm{d}t}.$$

Show that for a single gas in the vacuum space the mass flow is given by

$$\frac{\mathrm{d}m_0}{\mathrm{d}t} = \frac{m_0}{kT}Q = \frac{M}{RT}Q.$$

(b) The conductance (C) of a pipe or orifice with a pressure drop ΔP is defined by

$$Q = C\Delta P$$

Find an expression for C for a hole in a thin plate of area A and for a long pipe. (c) The pump speed is defined for a gas flow Q into the vacuum space producing a pressure P by

$$S_0 = \frac{Q}{P}.$$

Show that, for a pump connected to a vacuum space by a conductance C, the effective pump speed is

$$S = \frac{CS_0}{C + S_0}.$$

(d) Show that the final pressure for a leak Q_i into the vacuum is Q_i/S.

9.11. From Exercise 9.10 it is clear that the optimum pump at low pressure is one which absorbs all molecules incident upon it. A surface cooled to 4.2 K by liquid helium will absorb all gases except He, H_2 and Ne. Consider a vacuum space of 1 m^3 with a cooled surface of area A at 4.2 K. Show that the pump speed for the surface is

$$S = A \sqrt{\frac{RT}{2\pi M}}$$

and evaluate S for nitrogen at room temperature when A is 10^{-2} m^2. What is the final pressure obtainable from such a pump if $Q_i = 10^{-4}$ $N\,m\,s^{-1}$?

9.12. Show that the peak intensity in the experiment to verify the Maxwell velocity distribution discussed in Section 9.3 occurs at

$$v = \sqrt{\frac{4kT}{m}}.$$

9.13. Estimate the coefficient of thermal conductivity (κ) and of viscosity (η) for argon molecules at NTP. Take the diameter of the molecules to be that calculated in Exercise 9.6. How are κ and η predicted to change if (a) the pressure is halved, (b) the temperature is halved?

9.14. One of the most important applications of molecular beams in ultra high vacuum is to the growth of integrated circuits (microchips). In the case of GaAs transistors for example the substrate is a single crystal of GaAs heated to about 850 K. At this temperature As will only remain on the substrate if it can combine with a Ga atom. The growth rate of the transistor therefore depends upon the flux of Ga atoms from the beam provided there is an excess of As atoms near the substrate. (A beam of Al atoms is also required to form the transistor but this need not concern us.) The substrate must be in an ultra high vacuum in order to reduce the impurity level in the GaAs to below 1 part in 10^8. The number of atoms or molecules of a particular species which stay on the substrate $(cm^{-2} s^{-1})$ may be written as a modification of eqn 8.25.

$$R = \alpha\xi P/(MT)^{\frac{1}{2}}$$

where α is a geometrical factor (given below for P in torr and M in kg), and ξ is the fraction of the incident molecules which remain on the surface. The saturated vapour pressure of Ga may be written

$$\log_{10} P(\text{torr}) = -\frac{11{,}020}{T} + 7\log_{10} T - 15.4$$

Assume each GaAs molecule occupies a volume of $64 \times 10^{-24} cm^3$. (a) Circuits are grown at a rate of 1 molecule layer per second. Show this is equivalent to a thickness of about 1 μm per hour. (b) Show that a flux of $6.25 \times 10^{14} cm^{-2} s^{-1}$ of Ga atoms is required to obtain 1 layer per s. (c) Show that the oven containing the Ga must be at 1236 K to obtain the flux in (b) if $\alpha = 8.8 \times 10^{19}$ and $\xi = 1$. (d) Show that the temperature of the oven must be stabilized to better than 0.5 K if the flux of Ga atoms is to vary less than 1%. (e) A typical gas impurity in the vacuum space is CO ($M = 28 \times 10^{-3}$ kg mol^{-1}) and α is 3.5×10^{22} for atoms coming from the vacuum. The best vacuum which can be obtained is about 5×10^{-11} torr. The impurity level in the GaAs film is found to be about 1 in 10^8. Show that this means that ξ must be about 10^{-4} for CO molecules on the heated substrate. A review is given in Luscher and Collins 1979.

PART IV

Applications of thermodynamics and statistical mechanics

10

Further applications

The basic principles of both the macroscopic (thermodynamic) approach and the microscopic (statistical mechanical) approach to systems in thermal equilibrium have been established in Parts I and II. In the present chapter a number of examples are given of the application of both approaches to systems which have features which are currently of interest in physics research.

There is a basic difference between statistical mechanics and thermodynamics which should be appreciated before one begins to apply the equations to real systems. The functional results of thermodynamics, the difference between C_P and C_V for example, are valid for any system in thermal equilibrium but the values of C_P and C_V individually, or of any other thermodynamic quantity, are unknown except at absolute zero. In Chapter 7 it was shown that statistical mechanics could provide a general expression for the specific heat at constant volume

$$C_V = k\beta^2 \left(\frac{\partial^2 \ln Z}{\partial \beta^2} \right)_V$$

where Z is the partition function and $\beta = kT^{-1}$. A knowledge of the partition function is in fact sufficient to derive all the thermal properties of a system with a conserved number of particles in thermal equilibrium. The difficulty which arises is that, with a few exceptions, it is not possible to evaluate an exact partition function for a real system. A model system is therefore chosen which should show the essential features of the real system and a partition function is derived for the model. This procedure is not of course peculiar to statistical mechanics but is the normal scientific approach to a complicated problem. One should always remember however that to say that a particular real system behaves as a 'perfect gas' or that the conduction electrons in a metal may be treated as a quantum gas (neglecting interactions) is to make additional and possibly incorrect assumptions which are not required by thermodynamics.

10.1 Quantum gases

In section 6.5 it was shown that the number of particles of a perfect gas in an energy level with degeneracy g_r was

$$n_r = \frac{g_r}{e^{\alpha + \beta \varepsilon_r} \pm 1}$$

(eqn 6.43) where the positive sign applies to fermions and the negative sign to bosons. The parameter β is as usual equal to $1/kT$ and α is defined by the equation

$$N = \sum_r \frac{g_r}{e^{\alpha + \beta \varepsilon_r} \pm 1} \tag{10.1}$$

where N is the total number of particles. (It is shown in Section 7.10 that $\alpha = -\beta \mu$, where μ is the chemical potential). This equation can always be turned into an integral over the density of states function for fermions but it will be seen later in this section that difficulties arise for bosons at low temperature. At high temperature however eqn 10.1 may always be written

$$N = 2\pi V g \left(\frac{2m}{h^2}\right)^{\frac{3}{2}} \int_0^\infty \frac{\varepsilon^{\frac{1}{2}} \, d\varepsilon}{e^{\alpha + \beta \varepsilon} \pm 1}$$

(eqn 6.46) and the classical limit was seen in Section 6.5 to occur when exp $\alpha \gg 1$. A gas of molecules at NTP is safely within the classical limit (see Exercise 6.7) but it is of some interest to find the first quantum correction term when exp α is large but the term ± 1 in the denominator of eqn 6.46 cannot be ignored completely. A gas under these conditions is said to be slightly degenerate.

The mean energy of the gas may be written

$$\bar{E} = N\bar{\varepsilon} = 2\pi V g \left(\frac{2m}{h^2}\right)^{\frac{3}{2}} \int_0^\infty \frac{\varepsilon^{\frac{3}{2}} \, d\varepsilon}{e^{\alpha + \beta \varepsilon} \pm 1}. \tag{10.2}$$

The integral cannot be evaluated in closed form, but for exp α much greater or less than unity, it may be expressed in the form of an infinite series. Only the first correction term is in fact of interest for a slightly degenerate gas and is found to be (Huang, 1963)

$$\bar{E} = \tfrac{3}{2} NkT \left(1 \pm \frac{h^3 N}{2^{\frac{5}{2}} g (2\pi m kT)^{\frac{3}{2}} V} - \cdots \right) \tag{10.3}$$

where the positive sign refers to fermions and the negative sign to bosons.

Since the mean energy of a classical perfect gas is $3NkT/2$ (eqn 8.18) the effect of the Pauli exclusion principle is to increase the mean energy of a gas of fermions. The free particles therefore behave as if there were a repulsive interaction between them. Similarly bosons have a lower mean energy than a gas of classical particles which is equivalent to an effective attraction between the particles. The classical perfect gas was defined to have a mean energy independent of its volume but it is clear from eqn 10.4 that the mean energy of the quantum perfect gas does depend upon the volume, due to the effective interaction between the particles introduced by quantum mechanics.

The correction term to the classical mean energy shown in eqn 10.3 is in fact quite negligible for even the lightest molecules. At NTP the value of the second term in the bracket for molecular hydrogen (a boson gas), for example is $\approx 10^{-6}$. The mean energy of the gas may be related to the pressure exerted by the gas using the relationship

$$\bar{E} = \tfrac{3}{2} PV \tag{10.4}$$

which has already been proved for a classical gas (eqn 7.67). The extension of eqn 10.4 to quantum gases follows because the volume dependence of the energy of a given energy level is (Appendix IV)

$$\varepsilon_n = \frac{h^2 n^2}{8\pi^2 m V^{\frac{2}{3}}}. \tag{10.5}$$

It may be shown from quantum mechanics that a particle in the state n will remain in that state if the volume is changed sufficiently slowly under adiabatic conditions. The change in the mean energy with volume is related to the pressure by the thermodynamic equation

$$P = -\left(\frac{\partial \bar{E}}{\partial V}\right)_S \tag{10.6}$$

and if the number of particles in each state is unchanged the right-hand side of eqn 10.6 is equal to $-2E/3V$ from which eqn 10.4 follows immediately.

The pressure exerted by a slightly degenerate gas may therefore be written

$$PV = NkT\left(1 \pm \frac{Nh^3}{2^{\frac{5}{2}} g (2\pi mkT)^{\frac{3}{2}} V}\right) \tag{10.7}$$

where the positive sign applies to fermions and the negative sign to bosons.

Although this equation has the form of the virial expansion introduced in Section 2.2. it should be clear from the earlier discussion of the magnitude of the correction term that this quantum correction is not of practical importance for real gases.

Equation 10.7 may be written

$$PV = NkT\left(1 \pm \frac{\lambda^3 N}{2^{\frac{5}{2}} V}\right) \tag{10.8}$$

where

$$\lambda = \frac{h}{\sqrt{(2\pi mkT)}} \tag{10.9}$$

is (apart from a small numerical factor) the de Broglie wavelength of a typical particle at temperature T (eqn 6.50). The correction term therefore depends upon the ratio of the de Broglie wavelength to the mean separation of the particles. While this quantum correction is always negligible (relative to the effect of particle–particle interactions in a real gas) a quantum correction to the virial coefficient calculated by classical physics is necessary when the de Broglie wavelength becomes of the order of magnitude of the dimensions of a molecule, as is possible for hydrogen and helium.

It might appear from the discussion up to this point that the theory of the quantum gases was of little practical importance. While it is certainly true that the effects discussed so far are quite negligible for gases there are systems which may in some sense be treated as weakly interacting to which the quantum gas calculations apply at least approximately. (Bose–Einstein statistics also apply to radiation (photons) as will be discussed in the next section.)

The conditions under which quantum effects will become pronounced (a degenerate gas) are clearly just the converse of the conditions for the classical limit. When the particles have small mass, the system is at low temperature and the number density of particles is large, the classical approximation ceases to be satisfactory. It does not however follow that the resulting system will be a perfect quantum gas since the combination of low temperature and high number density is likely to lead to strong interactions between the particles. The neglect of these interactions must therefore be justified for any system to which the quantum gas calculations are applied. It is perhaps worth repeating that the expression 'low temperature' applied to quantum systems does not necessarily mean a small number on the absolute scale of temperature but rather that the system is at a temperature much lower than some characteristic temperature for the system (Section 6.3). The electrons in a white dwarf star for example at a temperature of 10^7 K will be seen later in this section to be in the low-temperature limit.

The first application of Fermi–Dirac statistics to real systems was to the conduction electrons in metals. Drude had earlier given a classical theory of conductivity in metals which, although in some respects successful, contained features that were quite inexplicable within classical physics.

The outstanding feature of metals is their high electrical and thermal conductivity. Drude suggested that the valence electrons in a metal become detached from their parent atoms and are free to move through the crystal. These electrons are able to respond to a potential difference or a thermal gradient across the sample and are therefore called conduction electrons. Drude treated the conduction electrons as a Maxwell–Boltzmann gas and used the kinetic theory results for the thermal conductivity (eqn 9.16) and the electrical conductivity (eqn 9.42) leading to the expression

$$\frac{\kappa}{\sigma T} = 3\left(\frac{k}{e}\right)^2 \tag{10.10}$$

which is a constant for any metal. This remarkable result is in good agreement with experiment except at low temperature.

The concept of the free-electron gas must be taken seriously but, as was noted earlier in the discussion of the transport theory of gases (Chapter 9) the agreement found between theory and experiment for the ratio of two quantities is not necessarily a proof of the correctness of the details of the theory but may be due to fortuitous cancellations in the individual terms. The temperature dependence of the electrical conductivity of metals for example is in fact quite different from that predicted by eqn 9.42. A further difficulty of the classical theory of free electrons in metals is that the free electrons do not appear to contribute to the thermal capacity at room temperature. According to the equipartition theorem, N free electrons (one per atom in silver for example) have a thermal capacity of $3Nk/2$. The thermal capacity of a metal should therefore be greater than that of an insulator by this amount. At room temperature however there is little difference between the thermal capacities of metals and insulators.

Sommerfeld realised that even if the conduction electrons could be treated as a gas the condition for the classical limit (eqn 6.49) would not be satisfied due to the small mass of the electron and the high number density if each atom in the metal contributed one or more free electrons (Exercise 6.7). The electron gas in a metal is in fact in the extreme quantum limit and has a large mean energy even at the absolute zero of temperature, due to the effect of the Pauli exclusion principle which allows only one electron to occupy each single-particle state.

Equation 6.43 will be rewritten in the standard notation of solid state physics by substituting $-\beta\mu$ for α where μ is the chemical potential. Then at absolute zero the probability of occupancy of a single-particle state r of a gas of fermions

$$P_r = \frac{1}{e^{\beta(\varepsilon_r - \mu)} + 1} \tag{10.11}$$

is unity for all states with energy less than μ and zero for all states with energy greater than μ (Fig. 10.1). The value of μ at absolute zero is usually written ε_F (the Fermi energy). The Fermi energy may be found from eqn 6.46 in the form

$$N = 2\pi V g \left(\frac{2m}{h^2}\right)^{\frac{3}{2}} \int_0^{\varepsilon_F} \varepsilon^{\frac{1}{2}} \, d\varepsilon \tag{10.12}$$

since no states are occupied above ε_F. Then

$$\varepsilon_F = \frac{h^2}{8\pi^2 m} \left(\frac{3\pi^2 N}{V}\right)^{\frac{2}{3}} \tag{10.13}$$

where the factor of 2 has been substituted for g since electrons have a spin of 1/2.

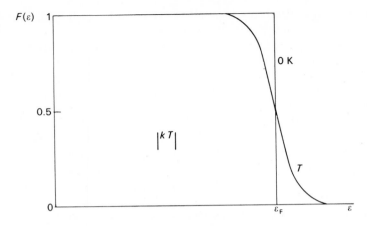

Fig. 10.1   The Fermi-Dirac distribution function at absolute zero and at temperature T ($T \ll T_F$). The energy range equivalent to kT is shown.

Similarly the mean energy of the gas at absolute zero is (from eqn 10.12)

$$\bar{E} = 2\pi V g \left(\frac{2m}{h^2}\right)^{\frac{3}{2}} \int_0^{\varepsilon_F} \varepsilon^{\frac{3}{2}} \, d\varepsilon \qquad (10.14)$$

$$= \tfrac{3}{5} N \varepsilon_F$$

The value of the Fermi energy of a metal found from eqn 10.13 assuming that each atom contributes one free electron is typically about 5 eV. A Fermi temperature may be defined as usual by the equation

$$\varepsilon_F = k T_F$$

and for 5 eV is approximately equal to 5×10^4 K. A normal laboratory temperature is therefore much less than the Fermi temperature and the mean energy of the electron gas at 0 K is equal to that of a classical gas at 2×10^4 K.

At temperatures above absolute zero the value of μ is no longer equal to the value ε_F but must be found from eqn 6.46 (where $\alpha = -\beta\mu$). When ε_F is 5 eV however the value of μ is found to decrease by only about 0.04 per cent of ε_F by room temperature. The main effect of the increase in temperature is that some electrons are now excited above the Fermi level (Fig. 10.1). The bulk of the electrons do not change their energy and so do not contribute to the specific heat.

The change in the populations of the energy states with temperature is shown in Fig. 10.2. The number of particles in a small range of energies $d\varepsilon$ is given by

$$dN = 2\pi V g \left(\frac{2m}{h^2}\right)^{\frac{3}{2}} \frac{\varepsilon^{\frac{1}{2}} \, d\varepsilon}{e^{\beta(\varepsilon - \mu)} + 1}. \qquad (10.15)$$

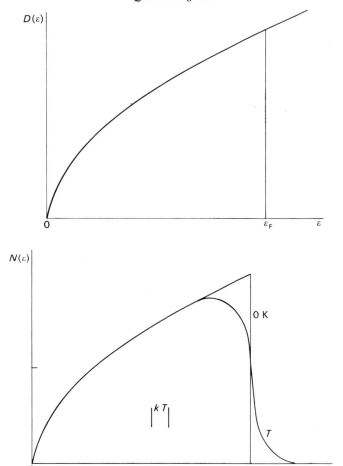

Fig. 10.2 (a) The form of the density of single-particle states as a function of energy for free electrons. (b) The number of electrons per unit energy range at absolute zero and at temperature T ($T \ll T_F$). The energy range equivalent to kT is shown.

At absolute zero there are no electrons above ε_F but at temperature T the electrons with energy within $\approx kT$ of the Fermi energy move into higher energy states. The bulk of the electrons cannot respond to the temperature change because there are no unoccupied states within the energy range kT if $T \ll T_F$.

The number of excited electrons is

$$N_{ex} \approx N \frac{T}{T_F}$$

and each excited electron has gained energy of order kT so

$$\Delta E \approx \frac{NkT^2}{T_F}$$

and the heat capacity due to the free electrons is

$$C_{el} \approx \frac{2NkT}{T_F} \qquad (T \ll T_F). \tag{10.16}$$

This expression is almost correct. The rigorous calculation using eqn 6.46 to find μ as a function of temperature leads to

$$C_{el} = \frac{\pi^2 NkT}{2T_F} \qquad (T \ll T_F) \tag{10.17}$$

for N free electrons. The classical theory would of course predict

$$C_{el} = \tfrac{3}{2} Nk \quad \text{(classical)} \tag{10.18}$$

which is some fifty times greater at room temperature. The electronic contribution to the specific heat at room temperature is therefore negligible relative to the lattice specific heat which is of order $3Nk$ since at high temperature the lattice vibrations may be treated as independent three-dimensional simple harmonic oscillators (Section 10.3). The quantum theory of the electron gas therefore explains the main discrepancy of the classical theory of the free electrons in metals. At low temperature the electron contribution to the specific heat may be observed because it decreases more slowly with temperature than the lattice contribution (see Figs. 4.2 and 10.9).

The quantum theory also reproduces the most striking success of the classical theory (eqn 10.10) with a slightly different numerical coefficient.

$$\frac{\kappa}{\sigma T} = L = \frac{\pi^2}{3} \left(\frac{k}{e}\right)^2 \tag{10.19}$$

where L is called the Lorentz number and is equal to 2.5 $\times 10^{-8}$ watt ohm K^{-2}. This equation, as has already been remarked, is successful near room temperature but as in the kinetic theory discussion of Chapter 9 it is clear that a ratio of this type involves identical mean free paths for both processes. At low temperature this ceases to be true and L may decrease by up to an order of magnitude.

One further application of the Fermi–Dirac statistics will be briefly discussed in this section. The Fermi energy of a metal has been seen to be typically 5 eV, equivalent to a Fermi temperature of 5×10^4 K. A normal laboratory experiment is therefore performed at a temperature $T \ll T_F$ and indeed the melting point of silver for example is only $\approx 10^3$ K. Since the magnitude of the Fermi temperature depends upon the number density of electrons in the system (eqn 10.13) a star may be in the low-temperature

($T \ll T_F$) limit even though the temperature is extremely high ($\approx 10^7$ K) if the density of matter contained in it is much greater than normal. The theory of the Fermi–Dirac gas has been applied to a class of stars known as white dwarf stars which are believed to consist almost entirely of helium. The temperature of the centre of the star ($\approx 10^7$ K) is sufficient to ionise the helium atoms and the density of the helium is such that the concentration of free electrons is about 10^{36} m^{-3}, corresponding to a Fermi temperature of 10^{11} K. The electron gas is therefore in the degenerate limit $T \ll T_F$. The astronomical problem is however more complicated than that of the metal, since the electrons are moving sufficiently rapidly to have to be treated by relativistic mechanics. The equilibrium of the star also requires that the pressure exerted by the Fermi gas (Exercise 10.2) be balanced by the gravitational attraction of the helium nuclei. The calculation of the mass–radius relationship for a white dwarf star will not be performed here [see Huang (1963) or Kittel and Kroemer (1980)] but leads to the conclusion that the mass of a star cannot be greater than 1.4 times the mass of the sun if it is to become a white dwarf. A greater mass exerts such a large gravitational attraction that it cannot be balanced by the pressure exerted by the Fermi gas. This prediction appears to be in agreement with astronomical observation and represents a striking example of the application of quantum gas theory to a system which could hardly be further from our original picture of a gas as a widely separated set of particles.

The justification for the application of quantum gas theory to dense systems of rapidly moving charged particles such as electrons is really part of solid-state physics and will not be considered in detail here, but the physical idea behind the theory is fairly easy to see. The electrons are free to move against a background of ionized atoms so the long-range Coulomb interaction between two bare electrons is screened by the positive background and the interaction is then found to depend exponentially on distance. The efficiency of the screening increases with the density of particles so the importance of interactions between the electrons (the ratio of potential to kinetic energy) actually decreases with increasing density. A dense gas of electrons (with an equal background charge of opposite sign) is a more perfect gas than a similar dilute gas.

An electron–electron scattering process is also restricted by the exclusion principle. The final electron states must be unoccupied if the scattering process is to occur, so only electrons with an energy within about kT of the Fermi level are affected. The effect of electron–electron interactions can therefore be ignored, to a first approximation. It is also easy to show that an electron is not scattered at all by a regular lattice so the electron–lattice scattering (leading to a finite electrical conductivity) is a weak interaction due only to the small-amplitude vibrations of the ions about their equilibrium positions. The mean free path of an electron in a pure metal is therefore expected to be much greater than a lattice spacing and this is supported by experiment. The mean free path of an electron in copper at room temperature is found to be about one hundred lattice spacings and increases at lower temperatures.

The *Bose–Einstein distribution* has also been applied to a system which is very different from a gas. Helium is the only substance not to solidify under atmospheric pressure at low temperature and it is believed that even at absolute zero it would continue to exist as a liquid under conditions of saturated vapour pressure. A pressure of at least 25 atmosphere is required to solidify liquid helium. Helium liquefies at 4.2 K at atmospheric pressure and behaves as a normal fluid of low viscosity until a temperature of 2.17 K. At lower temperature the properties of liquid helium are unique. Liquid helium is often called helium I in the normal region and helium II in the temperature region below 2.17 K. Helium II appears to have an enormously high thermal conductivity and the value of its coefficient of viscosity depends upon the method of measurement. A Poiseuille flow experiment using tubes of small diameter suggests that the coefficient of viscosity is zero but the damping of the oscillations of a pendulum leads to a finite value for the viscosity. A partial explanation of these properties (due to Tisza) considers helium I to be a normal fluid but helium II to be a mixture of normal fluid and superfluid. The concentrations of the two fluids are a function of temperature. The superfluid has zero viscosity and therefore determines the flow in small tubes but the normal component which also exists at finite temperature damps the motion of a pendulum. Helium II has many other extraordinary properties which will not be considered here.

Natural helium is composed almost entirely of the isotope ^{4}He which is a boson system. The rare isotope of helium ^{3}He is a fermion and liquid ^{3}He does not show a transition at 2.17 K. It is natural to conclude that the superfluid properties of helium II are connected with the statistics of ^{4}He. (Recent research however has shown that ^{3}He is also a superfluid at temperatures below 3 mK.) There is less justification for treating liquid helium as a gas of bosons than there was for applying the theory of the Fermi gas to the free electrons in metals, since interactions between the atoms must be of importance, but an approximate theory of helium II was developed by London following Einstein's calculation for the quantum gas.

Einstein noted that it was possible in certain circumstances for a large fraction of the particles of a boson gas to condense in the lowest energy level of the system. London suggested that the fraction of the particles in the lowest energy level corresponded to the superfluid component of helium II, and the particles in the upper states to the normal components. The difference between ^{4}He and ^{3}He at low temperature would therefore follow since only one atom of ^{3}He could occupy the lowest state, due to the exclusion principle for fermions.

The possibility of the condensation of a significant fraction of the molecules of a boson gas into the ground state will now be examined. The number of bosons in energy level r of degeneracy g_r may be written (once again substituting $\alpha = -\beta\mu$)

$$n_r = \frac{g_r}{e^{\beta(\varepsilon_r - \mu)} - 1} \qquad (10.20)$$

Now if the state of lowest energy (ε_0) is taken to have an energy equal to zero it is apparent that μ must always be less than zero if n_0 is to be positive. The value of μ depends of course (if other factors are kept constant) on the temperature and at sufficiently high temperature will be large and negative (Section 6.5). At some temperature it is possible that, for a given system, μ will approach the ground state energy of zero (from below) (Fig. 10.3). The number of particles in the ground state would then become much larger than that in the first excited state and would form a significant fraction of the particles in the system. The ratio of the numbers of bosons in the ground state (n_0) and the first excited state (n_1) is (from eqn 10.20)

$$\frac{n_0}{n_1} = \frac{e^{\beta(\varepsilon_1 - \mu)} - 1}{e^{-\beta\mu} - 1}$$

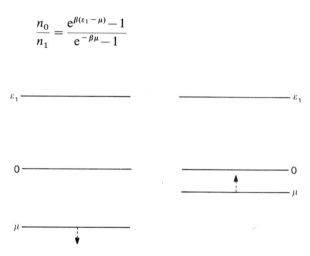

Fig. 10.3 In a degenerate gas of bosons the value of μ may approach that of the ground state (from below) so that $|\mu|$ is much less than the energy of the first excited state (ε_1). In the classical limit μ goes to minus infinity.

and when μ is small (and negative)

$$\frac{n_0}{n_1} = -\frac{\varepsilon_1}{\mu}$$

so n_0 may become much greater than n_1.

A comparison with the behaviour of the classical gas may help to show the unique nature of the boson gas. The populations of the ground and first excited states of the classical gas are given by

$$n_0 = n_1 e^{\beta \varepsilon_1}$$

Now ε_1 may be written in temperature units as kT_1 and has been shown to be of the order of 10^{-14} K (Exercise 6.3). At all temperatures accessible to experiment therefore

$$n_0 \approx n_1 \left(1 + \frac{T_1}{T}\right) \approx n_1$$

so the number of particles in the ground state is effectively equal to the number in the first (and many higher) excited states.

The ground state of a boson system must therefore be treated with some care. The normal conversion of the sum (eqn 10.1) into an integral involves the use of the density of states function

$$D(\varepsilon) = 2\pi V g \left(\frac{2m}{h^2}\right)^{\frac{3}{2}} \varepsilon^{\frac{1}{2}} \, d\varepsilon$$

which excludes the ground state ($\varepsilon = 0$) altogether. This approximation is always valid for fermions since the ground state can contain only one particle (apart from spin) and is also of no importance in the classical limit or the case of the slightly degenerate boson gas, where there are many states with effectively the same number of particles as the ground state, but must be improved upon for a degenerate gas of bosons. The simplest (and in fact adequate) approach is to use the normal density of states function, but to specify the ground state separately. If the number of particles in the ground state is N_0 and the number in all the excited states N_1 then the total number N is given by

$$N = N_0 + N_1$$

$$= \frac{1}{e^{-\beta \mu} - 1} + 2\pi V g \left(\frac{2m}{h^2}\right)^{\frac{3}{2}} \int_0^\infty \frac{\varepsilon^{\frac{1}{2}} \, d\varepsilon}{e^{(\varepsilon - \mu)\beta} - 1}.$$

At sufficiently high temperature we have already seen that μ is large and negative. In the high temperature limit therefore N_1 is effectively equal to N. As the temperature is lowered (keeping the volume constant) it is clear from inspection of the integral that μ becomes less negative to keep N_1 equal to N. At some temperature T_c, N_1 can only be equal to N if μ is equal to zero. The temperature T_c may be found from the equation

$$N = 2\pi V g \left(\frac{2m}{h^2}\right)^{\frac{3}{2}} \int_0^\infty \frac{\varepsilon^{\frac{1}{2}} \, d\varepsilon}{e^{\frac{\varepsilon}{kT_c}} - 1}.$$

The integral may be expressed in the form of a standard integral, eqn A2.9, to give

$$N = 2.61 \ V \left(\frac{2\pi mk T_c}{h^2} \right)^{\frac{3}{2}}$$

$$T_c = \left(\frac{N}{2.61 \ V} \right)^{\frac{2}{3}} \left(\frac{h^2}{2\pi mk} \right). \tag{10.21}$$

The critical temperature therefore occurs when the average particle separation is of the order of the average de Broglie wavelength of the particles.

Now in fact μ cannot be equal to zero but must be very small and negative at all temperatures equal to or less than T_c. The number of particles in the excited states at temperature T is therefore

$$N_1 = N \left(\frac{T}{T_c} \right)^{\frac{3}{2}} \qquad (T < T_c)$$

and the fraction in the ground state

$$\frac{N_0}{N} = 1 - \left(\frac{T}{T_c} \right)^{\frac{3}{2}}.$$

The fraction of the particles of a boson gas in the ground state is therefore negligible above T_c and increases at lower temperature as shown in Fig. 10.4. Since the ground state has zero energy and zero momentum the behaviour of a boson gas below T_c is sometimes described as a condensation in momentum space by analogy with the liquid–vapour transition where the liquid condenses in real space.

The critical temperature of a boson gas calculated from eqn 10.21 using the density of liquid helium ($145 \ \mathrm{kg \, m}^{-3}$) is $3.0 \ \mathrm{K}$, which is in quite good agreement with the experimental value of $2.17 \ \mathrm{K}$. The basic idea of the London theory of helium II, relating the superfluid component to the particles in the ground state, therefore seems to be valid and has been retained in more recent work, but the results of the theory of the boson gas are often in complete disagreement with the experimental results for liquid helium II.

The temperature T_c for a gas of bosons (eqn 10.21) increases with the particle density and therefore with the pressure exerted on the system, but the superfluid transition in helium II decreases with pressure. The behaviour of the specific heat of helium II close to the superfluid transition (Fig. 10.5) is also not in agreement with the specific heat of a gas of bosons which varies as $T^{3/2}$ below T_c. The transition temperature in helium II is often called the lambda point (λ), from the shape of Fig. 10.5. The transition of the boson gas is also fundamentally different from that in helium II in that it involves a latent heat (a first-order phase transition, Section 10.4) but there appears to

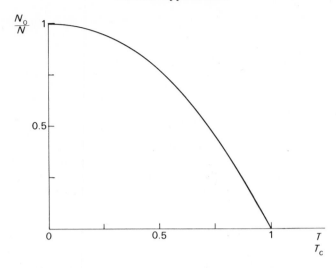

Fig. 10.4 The fraction of the particles in the ground state of a boson gas as a function of temperature. The fraction of the particles in the ground state of liquid ^{4}He appears to be only about 0.11 in the low temperature limit, rather than 1.0 (see e.g. Tilley and Tilley 1986).

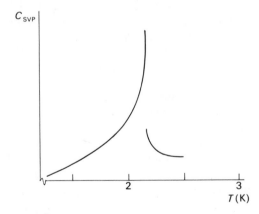

Fig. 10.5 The specific heat of liquid helium under conditions of saturated vapour pressure (SVP) shows a peak at 2.18 K (the lambda point). There is no latent heat associated with the peak.

be no latent heat associated with the lambda point. A better theory of helium II must therefore consider the theory of a quantum fluid rather than a quantum gas.

The theory of a degenerate boson gas does not at present apply to any real atomic system although it has obviously been of great importance for the qualitative understanding of helium II, but attempts are now being made to reach the critical density and temperature for Bose condensation in atomic hydrogen. At a temperature of 0.016 K, which is easily attainable, a density of 10^{24} atoms m^{-3} is required and the problem is of course to prevent the hydrogen atoms combining to form molecules. Details of the experimental arrangement will be found in Silvera and Walraven (1982).

Finally, we note that superfluidity can occur in Fermi systems such as the electrons in metals, where it is called superconductivity, or liquid ^{3}He. The essential requirement for superfluidity in a Fermi system is a coupling between the fermions to produce a quasi-particle with integer spin i.e. a boson. In metals the coupling between pairs of electrons is due to the interaction between the electrons and lattice vibrations (electron-phonon interaction). A strong electron-phonon interaction leads to rather poor electrical conductivity in the normal metal but to a high superconducting transition temperature, T_c. Until 1986 the highest T_c known was about 23 K (Nb$_3$Ge) but new materials such as YBa$_2$Cu$_3$O$_{7-\delta}$ have now been found which have T_c above 90 K.

10.2 Black body radiation

The experimental results for the behaviour of black body radiation had been established by Stefan before 1880 but a satisfactory microscopic theory was not available until Planck introduced the quantum hypothesis in 1900. During the period 1880–1900 however it became possible to understand many of the properties of black body radiation by purely thermodynamic arguments since thermodynamics does not require a microscopic theory for the system of interest. The present section therefore provides a good example of the usefulness of thermodynamics in the study of a system which is outside the scope of statistical mechanics at a given time. Another good example (which will not be considered in this book) is the superconductivity of certain metals at low temperature. The first experimental observation of super-conductivity was made by Kammerlingh Onnes in 1911 and thermodynamics had to be used to clarify the nature of the transition before a microscopic theory of the effect was developed by Bardeen, Cooper and Schrieffer in 1957.

The properties of the electromagnetic radiation in thermodynamic equilibrium with matter at temperature T are most easily studied by making a small hole in one side of a closed container. A sufficiently small hole will not disturb the equilibrium within the container (as in the case of gas effusion, Section 8.5) and the emitted radiation may be studied as a function of frequency and temperature. Radiation is often described as being 'at temperature T' although strictly the temperature refers to the temperature of the matter in equilibrium with the radiation. The radiation emitted from a small hole in a closed container is described as black body radiation since all radiation incident on the hole from outside the container is absorbed and the hole therefore acts as a perfect absorber or black body. Black body radiation contains the maximum possible amount of energy at every frequency for a given temperature because (as was first shown by Kirchoff) a perfect absorber is also a perfect emitter of radiation.

The experiments of Stefan showed that the total energy of black body radiation emitted per unit area per unit time at temperature T was of the form

$$R = \sigma T^4 \tag{10.22}$$

where T is the temperature on the thermodynamic scale and σ is a constant (the Stefan constant, equal to 5.67×10^{-8} watts $m^{-2} K^{-4}$). This result was then derived using thermodynamics by Boltzmann, and eqn 10.22 is often called the Stefan–Boltzmann law of black body radiation. The derivation of eqn 10.22 is quite straightforward from the energy equation (eqn 3.14)

$$\left(\frac{\partial U}{\partial V}\right)_T = T \left(\frac{\partial P}{\partial T}\right)_V - P.$$

The relationship between the pressure and the internal energy of black body radiation may be written

$$PV = \frac{1}{3} U$$

or

$$P = \frac{1}{3} u \tag{10.23}$$

where u is the energy density. This equation may be derived either from classical electrodynamics or by treating the radiation as a gas of photons. The energy density of black body radiation can only be a function of temperature, so eqn 3.14 becomes

$$u = \frac{T}{3} \frac{du}{dT} - \frac{u}{3}.$$

Therefore

$$u = \frac{T}{4} \frac{du}{dT}$$

and

$$u = bT^4 \tag{10.24}$$

where b is a constant. This equation is still not in the form required for a direct comparison with experiment since eqn 10.22 refers to the radiation emitted from a hole in the container, not to the energy density inside it. The simplest way to establish a relationship between the two equations is again to consider a gas of photons and to employ the argument used for the effusion of a gas through a small hole. Since all photons travel with the speed of light the number passing through unit area per unit time is simply $nc/4$ where n is the number of photons per unit volume. If the mean energy of a photon is $\bar\varepsilon$ then

$$u = n\bar\varepsilon$$

$$R = \frac{nc}{4}\bar\varepsilon = \frac{uc}{4} = \frac{bcT^4}{4}.$$

Therefore

$$\sigma = \frac{bc}{4}. \tag{10.25}$$

The temperature variation of the total energy of the total energy of black body radiation has therefore been derived by a thermodynamic argument, but the value of the constant cannot be found by this method.

The equation of state of black body radiation may now be written

$$P = \frac{u}{3} = \frac{b}{3}T^4 \tag{10.26}$$

showing that the pressure is independent of the volume. The heat capacity of black body radiation at constant pressure is infinite since an input of heat does not lead to an increase in temperature but the heat capacity of unit volume of radiation at constant volume (Exercise 10.3) is very small relative to that of matter.

The equation for a reversible adiabatic (isentropic) change will be seen later in this section to be of great importance. The required equation follows from the standard thermodynamic equation (3.15)

$$T\,\mathrm{d}S = \left(\frac{\partial U}{\partial T}\right)_V \mathrm{d}T + T\left(\frac{\partial P}{\partial T}\right)_V \mathrm{d}V$$

$$= \frac{\partial}{\partial T}\left(bVT^4\right)_V \mathrm{d}T + T\frac{\partial}{\partial T}\left(\frac{b}{3}T^4\right)_V \mathrm{d}V$$

$$= 4bVT^3\,\mathrm{d}T + \frac{4}{3}bT^4\,\mathrm{d}V.$$

In an isentropic process $\mathrm{d}S$ is zero and therefore

$$V\mathrm{d}T = -\frac{T}{3}\mathrm{d}V$$

or

$$VT^3 = \text{constant} \tag{10.27}$$

is the equation for an isentropic change of black body radiation.

This simple equation has been of importance firstly in the early attempt by Wien to understand the distribution of the energy of black body radiation as a function of temperature (which will now be described) and more recently in the controversy concerning the origin of the universe. The eqn 10.27 was used by Gamov to show that if the universe had begun as an extremely dense and hot concentration of matter and radiation (the primaeval fireball) which rapidly expanded (the big bang theory) then all space should be filled with black body radiation. The detection of this radiation (which will be discussed at the end of this section) has been one of the most interesting developments in astronomy in recent years.

The distribution of the energy of black body radiation at temperature T as a function of frequency may be described by a function $u(v,T)$ such that

> $u(v, T) \, dv$ is the energy density of black body radiation in the frequency range v to $v + dv$.

Therefore

$$u(T) = \int_0^\infty u(v, T) \, dv. \tag{10.28}$$

The experimental form of $u(v, T)$ is shown in Fig. 10.6. At a given temperature the function has a maximum value at some frequency (v_{max}) and decreases to zero at both limits. The value of v_{max} is seen to be proportional to the absolute temperature. The area under the curve for temperatue T represents the total energy density and is therefore proportional to T^4 (from eqn 10.22).

The Stefan–Boltzmann law was derived (apart from the value of the constant) without any assumptions about the detailed nature of radiation but thermodynamics is not sufficient to derive the equation of the isotherms shown in Fig. 10.6. Wien however did extend the thermodynamic argument to show that the isotherms must have the form

$$u(v, T) \, dv = v^3 f\left(\frac{v}{T}\right) dv \tag{10.29}$$

$$= T^4 f(x) dx \tag{10.30}$$

where x is equal to v/T. The Stefan–Boltzmann law therefore follows immediately

$$u(T) = \int_0^\infty u(v, T) dv = T^4 \int_0^\infty f(x) dx \tag{10.31}$$

since the integral over x is an (unknown) constant. Wien's argument will not be given in full here but, briefly, depends upon the calculation of the work done in an isentropic expansion by radiation pressure on a movable mirror. The radiation reflected from the mirror is Doppler shifted so that energy incident in the frequency range v to $v + dv$ is reflected into a new frequency range v' to $v' + dv'$. The calculation of the relationship between v and v' and the use of eqn 10.27 for an isentropic process then leads to eqn 10.29.

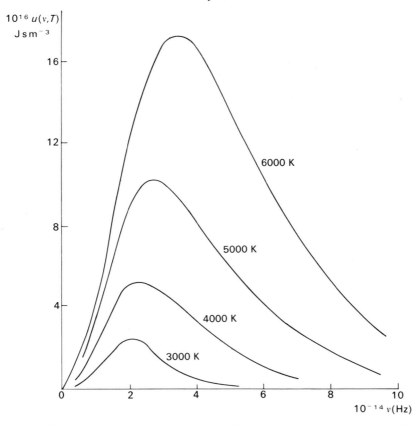

Fig. 10.6 The energy density of black body radiation at temperature T as a function of frequency.

The maximum in each isotherm shown in Fig. 10.6 will, according to eqn 10.29, have the form

$$\left[\frac{\partial u\,(v,\,T)}{\partial v}\right]_T = \frac{v^3}{T}f'\left(\frac{v}{T}\right) + 3v^2f\left(\frac{v}{T}\right) = 0.$$

Therefore

$$\frac{v_{\max}}{T}f'\left(\frac{v_{\max}}{T}\right) + 3f\left(\frac{v_{\max}}{T}\right) = 0.$$

The solution to this equation (assuming that it exists) must be of the form

$$\frac{v_{\max}}{T} = \text{constant} \tag{10.32}$$

in agreement with experiment. The equation of the maxima could also have been obtained by considering the energy density as a function of wavelength rather than frequency (Fig. 10.7) when

$$u(T) = \int_0^\infty u(\lambda, T)\,d\lambda \qquad (10.33)$$

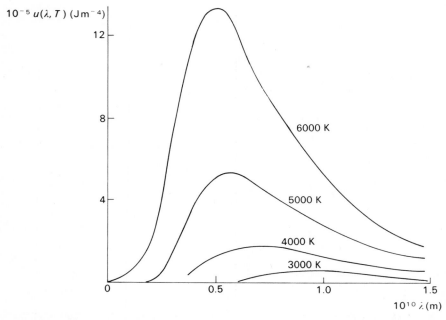

Fig. 10.7 The energy density of black body radiation at temperature T as a function of wavelength.

The form of $u(\lambda, T)$ may be derived from eqn 10.29 (Exercise 10.4)

$$u(\lambda, T) = \frac{c^4}{\lambda^5} f\left(\frac{c}{\lambda T}\right) \qquad (10.34)$$

(where λ is the wavelength and c the velocity of light) and

$$\lambda_{max} T = \text{constant.} \qquad (10.35)$$

This equation is usually called the Wien displacment law since it relates the position of the peak intensity (as a function of wavelength) to the temperature. It should be noted that the position of the maximum intensity measured as a function of wavelength is *not* the same as when measured as a function of frequency, that is to say $c \neq v_{max}\lambda_{max}$ (see Exercise 10.4).

Wien's law contains all the information about black body radiation which can be obtained from thermodynamics. Further progress therefore requires the use of statistical mechanics. However the classical statistical result (due to Rayleigh and Jeans) although giving excellent agreement with experiment at low frequency did not show a peak in the radiation isotherms at high frequency, but diverged (an effect sometimes called the ultraviolet catastrophe). The failure of classical statistical mechanics finally led Planck to introduce the concept of energy as a quantized (rather than continuous) function.

The Planck distribution law for $u(v, T)$ may be derived in two rather different ways due to Planck and Einstein. In the original classical calculation due to Rayleigh and Jeans the number of allowed modes of the electromagnetic radiation in the cavity were enumerated (the density of states function, Appendix IV) and each mode was assumed to have the same mean energy as a linear simple harmonic oscillator. Then

$$U(v, T)dv = Vu(v, T)dv = D(v)dv\,\bar{\varepsilon}.$$

Therefore

$$u(v, T)dv = \left(\frac{8\pi v^2}{c^3}dv\right)\bar{\varepsilon} \tag{10.36}$$

using eqn A4.23, multiplied by two for the two transverse polarizations of an electromagnetic wave, and

$$v = \frac{c}{\lambda} = \frac{ck}{2\pi}.$$

The mean energy of a linear simple harmonic oscillator at temperature T, according to classical physics, is given by the equipartition theorem to be kT (Section 7.6) so

$$u(v, T) = \frac{8\pi kT}{c^3}v^2\,dv \text{ (classical)}. \tag{10.37}$$

This expression contains no unknown constants and is in excellent agreement with experiment at low frequencies but *diverges* at high frequencies instead of going to zero.

Planck introduced the concept of the quantised sample harmonic oscillator with energy levels $\varepsilon = nh$ (where n is an integer and h a constant) to overcome this insuperable problem within classical physics. The mean energy of such an oscillator was calculated in Section 7.4

$$\bar{\varepsilon} = \frac{hv}{e^{\frac{hv}{kT}} - 1}$$

where the zero-point energy term $hv/2$ has been omitted since it leads to an infinite energy density at high frequencies. The Planck distribution law is therefore, substituting for $\bar{\varepsilon}$ in eqn 10.36

$$u(v, T)dv = \frac{8\pi h}{c^3} \frac{v^3\,dv}{e^{\frac{hv}{kT}} - 1}. \tag{10.38}$$

This expression is identical to the classical result in the limit $hv \ll kT$ but decreases to zero at high frequency. The frequency at which the distribution shows a maximum is given by

$$\frac{hv_{max}}{kT} = 3\left(1 - e^{-\frac{hv_{max}}{kT}}\right) \approx 2.82 \tag{10.39}$$

in agreement with the form of eqn 10.32. Equation 10.38 may also be written in the form

$$u(v, T) = v^3 f\left(\frac{v}{T}\right)$$

in agreement with the thermodynamic result due to Wien. The value of the Stefan constant may be found from

$$u(T) = \frac{8\pi h}{c^3} \int_0^\infty \frac{v^3\,dv}{e^{\frac{hv}{kT}} - 1}$$

$$= \frac{8\pi(kT)^4}{(ch)^3} \int_0^\infty \frac{x^3\,dx}{e^x - 1}$$

and the eqn 10.25 relating the energy density to the power radiated per unit area. Hence

$$\sigma = \frac{2\pi k^4}{c^2 h^3} \int_0^\infty \frac{x^3 dx}{e^x - 1}.$$

The definite integral is equal to $\pi^4/15$ (Appendix II) so

$$\sigma = \frac{2\pi^5 k^4}{15 c^2 h^3} \tag{10.40}$$

The value of h can be found from either eqn 10.39 or eqn 10.40 and is in excellent agreement with the value found for the Planck constant in other experiments. In fact the most accurate value for the Stefan constant is obtained not by direct experiment but by calculation using eqn 10.40.

The Planck radiation equation was of course the first calculation in physics which treated the energy of a system as quantized. The calculation given above may be felt to be rather uncomfortably close to the original classical argument since the only difference is in the treatment of the mean energy of a simple harmonic oscillator. The simple harmonic oscillator was chosen by Planck as the simplest example of a radiating body but the same result can be obtained by more elaborate quantum-mechanical calculations.

An alternative approach which has less connection with the classical argument is due to Einstein. In the theory of the photoelectric effect it is necessary to consider light not as continuous classical electromagnetic waves but rather as quanta of radiation (photons). The electromagnetic field in a container may be replaced by a gas of photons. The photons are indistinguishable particles with no restriction on the occupancy of energy levels (bosons) but unlike a gas of molecules the number of photons in the container is not constant. The Bose–Einstein distribution function has now to be calculated without the condition that led to the introduction of the undetermined multiplier α in Section 6.5. The number of photons in the level r of degeneracy g_r is therefore

$$n_r = \frac{g_r}{e^{\beta \varepsilon} - 1} \text{ (photons).} \tag{10.41}$$

The energy of a photon is related to the frequency by

$$\varepsilon = h\nu.$$

The energy density of states function for photons is determined by the relationship

$$\varepsilon = cp$$

(where p is the momentum), and the requirement that each photon state occupies a volume h^3 in phase space (Appendix IV). The density of states function is in fact identical with the density of states for the normal modes (eqn 10.36) so

$$u(\nu)\,d\nu = \left(\frac{8\pi\nu^2 d\nu}{c^3}\right)\left(\frac{h\nu}{e^{\beta h\nu} - 1}\right) \tag{10.42}$$

in agreement with eqn 10.38. The theory of black body radiation can therefore be considered from either a wave or a particle viewpoint. The development of the theory of black body radiation was far more complicated than has been described here. A brilliant historical account is given in Kuhn (1978).

The importance of the Planck theory of radiation in the history of atomic physics is too well known to be worth repeating here but the more recent use of the theory of radiation in cosmology and radioastronomy may not be so well known.

According to the 'big bang' theory the present universe began at some definite time as an enormously dense and hot cloud of matter and radiation which then expanded and cooled. Gamow pointed out in 1948 that if the big bang theory was correct the universe must be flooded with black body radiation which would look the same in all directions in space (isotropic radiathon). The detection of this radiation in 1964 was therefore a most important check on the theory of the evolution of the universe from an initial hot phase and seems to disprove the theories which require either a cold start to the universe or a 'steady-state universe' which has always looked as it does at present.

The basic concept of the Gamow theory is that the energy density of radiation (u) was greater than that of matter during the early development of the universe after the big bang. The proof is straightforward and depends only upon the cosmological postulate of uniformity which states that the universe may be considered to be divided into cells within all of which the conditions are identical. (The concentration of matter into galaxies in the universe at present must either be ignored or averaged over a sufficiently large volume.) The problem of the expanding universe therefore reduces to that of the properties of matter and radiation in a closed container undergoing a reversible adiabatic (isentropic) change of volume.

The present average density of matter in the universe is estimated to be 10^{-27} kg m^{-3} and of radiation 10^{-30} kg m^{-3}. The interaction between matter and radiation at present is extremely weak but we shall see that it must have been sufficiently strong at earlier stages in the evolution of the universe for the radiation to have been in thermal equilibrium with matter. The radiation must therefore be black body radiation.

The effect of decreasing our representative volume from its present value in an isentropic process is to increase the energy density, as may be seen from the eqns 10.24 and 10.27

$$u = b T^4$$

$$VT^3 = \text{constant.}$$

Therefore

$$u \propto V^{-4/3} \propto R^{-4} \tag{10.43}$$

where R^3 has been written for V. The rest energy of the matter of density ρ_m also increases but at a slower rate since

$$\rho_m c^2 \propto V^{-1} \propto R^{-3}. \tag{10.44}$$

The two densities were equal (using the estimates of the two densities at present) when the total density was 10^{-18} kg m^{-3} and the radiation density was greater than the matter density at earlier periods in the history of the universe. The history of the universe in this region will now be considered before returning to the properties of the radiation observed in the universe at present.

The behaviour of the universe during the first fraction of a second ($< 10^{-44}$ s) after the big bang can only be a matter for conjecture but at later times a plausible theory can be constructed based on nuclear physics and the general theory of relativity. After about 10^{-44} s the temperature of the universe was greater than 10^{12} K and the density was greater than 10^{17} kg m^{-3}. The universe consisted of high energy particles, antiparticles and photons. The average energy of a photon at temperature T_M is $\approx kT_M$ so photons can produce particle–antiparticle pairs if

$$kT_M \approx 2M_0c^2 \tag{10.45}$$

where M_0 is the rest mass of the particle. At temperature above 10^{12} K particles and antiparticles would be in equilibrium with photons but as the universe expanded and cooled below T_M, particles of rest mass M_0 would annihilate with their antiparticles and not be replaced by pair production. The presence of atomic particles in the present universe must be due to an initial small excess of particles over antiparticles after the big bang. It has been estimated that the present density of matter in the universe is consistent with an initial excess of particles over antiparticles of one part in 10^9 but it is not of course possible to account for this.

The temperature of the expanding universe is estimated to have dropped below 10^{12} K after 10^{-4} s. The heavy antiparticles would then all have annihilated but the antielectron (positron) continued to exist for about 10s by which time the temperature had reached 10^{10} K. The positrons then annihilated with electrons to form photons and left a small residue of electrons. Simultaneously nuclear reactions between protons and neutrons led to the formation of helium nuclei. The universe therefore consisted of photons and an ionised gas of protons, helium nuclei and electrons until the temperature had decreased to about 3×10^3 K some 10^6 years after the big bang. (The role of neutrinos will not be considered here.)

The electrons then combined with the ions to form a gas after which matter and radiation cooled separately since the photons could not interact with the gas. The equation for an isentropic change of black body radiation (10.27) may therefore be applied between a temperature of 3×10^3 K and the present. The radiation must be true black body radiation since it was initially in equilibrium with matter. The density of radiation and matter is estimated to be roughly equal at 3×10^3 K so the present temperature of the radiation may be found from eqn 10.24

$$u = bT^4$$
$$\therefore T \approx 3 \times 10^3 \left(\frac{10^{-30}}{10^{-18}}\right)^{\frac{1}{4}}$$
$$\approx 3\,\text{K}.$$

The original estimates of the temperature of the black body radiation by Gamow and his collaborators were in the range 25–5 K in surprisingly good agreement with the present best experimental value of 2.72 ± 0.04 K when the obvious uncertainties associated with the calculation are considered. Gamow was led to the prediction of the background radiation in 1948 while constructing a theory of the formation of the elements after the big bang. The theory was neglected however after it had been shown to be unable to explain the formation of elements beyond helium and the prediction of the background radiation was forgotten. Astronomers were not in fact equipped in 1948 to search for isotropic radiation which if it had a temperature of 3 K would have a maximum intensity in the microwave region near 10^{11} Hz.

The development of radioastronomy should have quickly led to the discovery of the background radiation and in fact it probably was observed but was attributed to electrical noise in the system. Dicke independently predicted the presence of isotropic black body radiation with a peak in the microwave region in 1964 but Penzias and Wilson made the first known observation of the background radiation before his apparatus had been completed.

Penzias and Wilson had designed a sensitive microwave system working at a wavelength of 7 cm to receive signals reflected from satellites. The electrical noise in the system was found to be above that which had been calculated and it was finally realised that a signal from the background radiation had been detected. The intensity of the background radiation has now been measured at a number of frequencies (Fig. 10.8) and is consistent with black body radiation with a temperature close to 3 K. The measurements above the peak at 10^{11} Hz which finally established the nature of the radiation were made in 1975 using balloon flights because the earth's atmosphere is strongly absor-

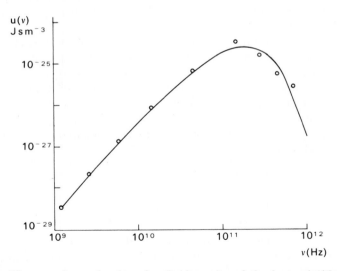

Fig. 10.8 The experimental values for the intensity of the isotropic black body background radiation as a function of frequency. The theoretical curve for a temperature of 2.99 K is also shown. The best value of the temperature, averaged over all measurements, is now 2.72 ± 0.04 K, (Witebsky *et al.* (1986).

bing at these frequencies. The lower limit to the measurements is set by the presence of strong galactic synchrotron radiation at frequencies near 10^8 Hz.

The intensity of the background radiation is therefore consistent with black body radiation of temperature $\simeq 3$ K and since the radiation is also found to be isotropic to the presently available accuracy of 0.1 per cent there seems little doubt that it constitutes good evidence for the big bang theory. A further improvement in the accuracy of the measurement of the intensity as a function of direction in space should show a small anisotropy due to the velocity of rotation of the solar system ($\simeq 200$km s^{-1}) about the centre of the galaxy.

The existence of black body radiation throughout the universe is of immense importance to astrophysics because the radiation density at 3 K is very large compared to all other energy terms except the rest energy density of matter. Peebles has estimated for example that if it were possible to ionize all the atoms in the universe at the expense of the background radiation the temperature of the radiation would decrease by only 10^{-5} K. The presence of the background radiation must therefore always be considered in astrophysical calculations.

10.3 Heat capacity of solids

The heat capacity at constant volume of a body was defined in Section 2.6

$$C_V = \left(\frac{\mathrm{d}Q_R}{\mathrm{d}T}\right)_V = \left(\frac{\partial \overline{E}}{\partial T}\right)_V = T\left(\frac{\partial S}{\partial T}\right)_V.$$

The heat capacity at constant pressure is normally obtained by experiment and then the correction discussed in Section 5.2 applied to reduce the results to constant volume. The heat capacity at low temperature is however of particular interest and in this region the difference between C_P and C_V can usually be neglected.

The heat capacity of a solid may be thought of as arising from a number of independent terms. All solids consist of an array of atoms which vibrate about their mean positions. Since the amplitude of the vibrations increases with temperature, there will be a change in the energy of the crystal lattice and a contribution to the heat capacity usually known as the lattice heat capacity. The contribution of the free (conduction) electrons in metals to the heat capacity has already been discussed in Section 10.1. In addition to these terms any other energy term of the solid which changes with temperature will contribute to the heat capacity. In this section only one such effect (the so-called Schottky effect) will be discussed in detail but a number of other examples are considered briefly.

The atoms of a crystalline solid form a regular array in space. The atoms vibrate about their mean positions with an amplitude which increases with temperature but even at the melting point the amplitude is much less than the interatomic spacing. An oscillation of small amplitude may always be represented, to a first approximation, by simple harmonic motion but the interaction between the atoms of a solid leads to a coupling between the N individual oscillators where N is the number of atoms in the crystal. The situation is similar to that discussed for the electronic states of a solid in Section 6.2. The crystal may be thought of as the bringing together of N oscillators each originally with the same frequency. As the oscillators begin to interact, the single frequency is changed into a range of frequencies which depend upon the strength of the interaction. The potential energy of one atom now depends upon the coordinates of all the other atoms in the crystal.

The N coupled oscillators can however always be transformed by the methods of mechanics (for simple harmonic motion) to a set of $3N$ uncoupled one-dimensional simple harmonic oscillators with certain allowed frequencies known as normal modes. The classical value for the lattice heat capacity therefore follows directly from the equipartition theorem. Since the mean energy of each simple harmonic oscillator is kT the lattice heat capacity is $3Nk$ or $3R$ for one mole of solid. This result was in fact first obtained by Dulong and Petit from experiments at room temperature. There are however exceptions to the rule (notably diamond) and the classical theory fails completely at low temperatures.

The quantum theory of the lattice heat capacity involves the behaviour of the quantized simple harmonic oscillator and is obviously similar to the discussion already given for black body radiation. There are however two features which in general distinguish the two cases and which arise from the discrete nature of the crystal lattice. Firstly since there are only $3N$ allowed normal lattice modes there must be a maximum lattice frequency and secondly the properties of a crystal may be different in different directions (anisotropy). These distinctions will be seen to be unimportant at low temperatures (because only oscillations of long wavelength are then possible) and in this limit a result equivalent to Stefan's law will be obtained.

The mean energy of a simple harmonic oscillator of frequency v_r was found in Section 7.4 to be (apart from the zero-point energy which is of no interest for the heat capacity)

$$\bar{\varepsilon}_r = \frac{hv_r}{e^{\beta hv_r} - 1}$$

so the lattice energy may be written

$$\bar{E} = \sum_{r=1}^{3N} \frac{hv_r}{e^{\beta hv_r} - 1} \tag{10.46}$$

where the v_r are the normal modes of the lattice.

It is worth noting that a 'particle' interpretation can be given to eqn 10.46. In analogy with the photon associated with the electromagnetic field the 'particles' are called phonons. The expression 'quasi-particle' is sometimes preferred for the phonons, to distinguish them from the atoms of the lattice. In the phonon picture there are n_r phonons with frequency v_r and the mean energy is

$$\bar{E} = \sum_r n_r h v_r$$

where

$$n_r = \frac{1}{e^{\beta h v_r} - 1}$$

The analogy between photons and phonons should now be clear. The phonons are bosons with no restriction on their number and therefore follow the same form of the Bose–Einstein statistics as photons.

The normal modes of the lattice may as usual be considered to be sufficiently close together to be approximated by a density of states function. The sum of eqn 10.46 may then be transformed into an integral

$$\bar{E} = \int_0^\infty \frac{hv}{e^{\beta h v} - 1} D(v) \, dv \tag{10.47}$$

where

$$\int_0^\infty D(v) \, dv = 3N \tag{10.48}$$

since the number of allowed modes must be restricted to $3N$. The statistical mechanics required for the lattice mean energy and therefore the lattice heat capacity is all contained in eqn 10.47 but the purely *mechanical* problem of evaluating the density of states function has still to be considered.

At sufficiently high temperature eqn 10.47 reduces to the classical result regardless of the form of the density of states function. The exponential term may be expanded to give $(1 + \beta hv)$ and

$$\bar{E} = kT \int_0^\infty D(v) \, dv = 3NkT. \tag{10.49}$$

At lower temperatures the form of the density of states function is more important but because it occurs inside an integral it will be seen that quite a crude model is sufficient to give reasonable agreement with measured lattice heat capacities. Conversely it is difficult to get any information about the density of states function from measurements of heat capacity.

The form of the density of states function $D(v)$ can be found for a given solid either by making some assumption about the number of neighbouring atoms each atom interacts with and then solving the resulting equations on a computer or by various experimental methods, notably neutron scattering (see Kittel (1986)). The density of states as a function of wave vector (Appendix IV) of a crystalline solid may be found from symmetry arguments (group theory) so the frequency of a phonon as a function of wave vector is required (the dispersion relation) before $D(v)$ can be evaluated. The measurement of the change of energy and momentum of a neutron beam scattered by the solid enables this relation to be established directly.

The first quantum calculation of the lattice heat capacity (due to Einstein) evaded the problem of the density of states function by treating the lattice as a set of N *independent* oscillators all with frequency v_E. The mean energy is therefore

$$\bar{E} = \frac{hv_E}{e^{\beta hv_E} - 1} \int_0^\infty D(v)\,\mathrm{d}v.$$

Therefore

$$\bar{E} = \frac{3Nhv_E}{e^{\beta hv_E} - 1}.$$

The lattice heat capacity is

$$C_V = \left(\frac{\partial \bar{E}}{\partial T}\right)_V = \left(\frac{\partial \bar{E}}{\partial \beta}\right)_V \frac{\mathrm{d}\beta}{\mathrm{d}T} = \frac{3Nk(\beta hv_E)^2 e^{\beta hv_E}}{(e^{\beta hv_E} - 1)^2} \tag{10.50}$$

where v_E is to be chosen to give the best agreement with experiment for a particular solid. An Einstein temperature is defined by the equation

$$hv_E = k\theta_E = \hbar\omega_E$$

and is found to be typically 200 K, corresponding to a frequency of about 5×10^{12} Hz which lies in the infrared region of the spectrum. The law of Dulong and Petit can now be understood since room temperature is greater than the Einstein temperature of most solids, and eqn 10.50 becomes

$$C_V \approx 3Nk \qquad (T \gg \theta_E).$$

The apparent exceptions to the law of Dulong and Petit arise for solids with an unusually high θ_E. Einstein showed that a reasonable fit to the measurements for diamond for example (which has a heat capacity about one quarter of the classical value at room temperature) was obtained using eqn 10.50 with θ_E equal to 1320 K.

At low temperature the Einstein equation is not usually in good agreement with experiment. The theory shows an exponential decrease of the heat capacity with temperature.

$$C_V = 3Nk\left(\frac{\theta_E}{T}\right)^2 e^{-\frac{\theta_E}{T}} \qquad (T \ll \theta_E) \tag{10.51}$$

but for most solids the lattice heat capacity is found to be proportional to T^3. This discrepancy is not surprising because at low temperature the low-frequency ($hv_r \approx kT$) normal modes of the solid are important and these have been replaced by one high-frequency mode in the Einstein model. There are solids however, in which one particular lattice mode is important, which are

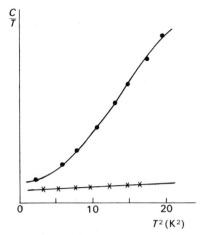

$T^2 (K^2)$

Fig. 10.9   An example of a solid whose heat capacity is well described by the Einstein model. The lattice vibrations of Al_{10} V (upper curve) contain one low-frequency mode which dominates the specific heat ($T_E = 22$ K). A normal metal is shown in the lower line (see also Fig. 4.2). (After Caplin, Grüner and Dunlop 1973)

quite well described by the Einstein model (Fig. 10.9) or a combination of the Einstein model and that of Debye discussed later in this section.

The theory of the low-temperature lattice heat capacity may be approached in a rather general manner because only low energy (long wavelength) modes can be excited. The structure of the lattice is therefore not important and the dispersion relation becomes simply

$$c_s = v\lambda = \frac{2\pi v}{k} \tag{10.52}$$

where c_s is the velocity of sound in the crystal and k the wave vector. In general c_s will be different for transverse and longitudinal waves and for different directions in the crystal, but these refinements will not be considered here.

The discrete lattice may be approximated to by an elastic continuum in the low temperature (long wavelength) region and the density of states function is then given by the same form as for black body radiation (Appendix IV)

$$D(k)\,dk = \frac{3V}{2\pi^2} k^2\,dk \tag{10.53}$$

where the factor of 3 allows for the two transverse and one longitudinal waves with wave vector k. This equation leads to an *infinite* number of allowed modes when integrated over all values of k, rather than the correct value of $3N$, but *at sufficiently low temperature* this is of no importance for the calculation of the mean energy of the lattice because the integral (eqn 10.47) is effectively cut off at high frequency by the exponential term.

Using eqn 10.52 the density of states as a function of frequency becomes

$$D(v)\,dv = \frac{12\pi V v^2\,dv}{c_s^3} \tag{10.54}$$

and the energy

$$\bar{E} = \frac{12\pi V}{c_s^3} \int_0^\infty \frac{h v^3\,dv}{e^{\beta h v} - 1}$$

$$= \frac{12\pi V}{c_s^3 h^3} (kT)^4 \int_0^\infty \frac{x^3\,dx}{e^x - 1} \quad (T_{\to 0\,\mathrm{K}}) \tag{10.55}$$

which is of the same form as Stefan's law (eqn 10.40). The lattice heat capacity is therefore proportional to T^3 *at sufficiently low temperature*; a result first obtained by Debye. The T^3 law in fact follows simply from the form of the density of states function in three dimensions. The general result for a system of n dimensions is T^n (Exercise 10.5).

The T^3 law is in good agreement with experiment for most insulators at sufficiently low temperature (Fig. 10.10) although highly anisotropic materials such as graphite which form in layers are closer to T^2 behaviour. Thin films also follow the T^2 law as would be expected for nearly two-dimensional

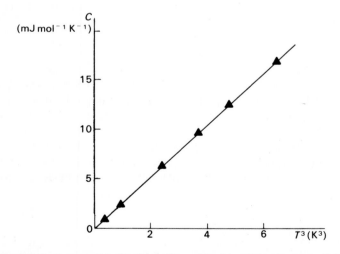

Fig. 10.10   The heat capacity of a three-dimensional insulator is proportional to the cube of the temperature at sufficiently low temperature. Values are shown for solid argon.

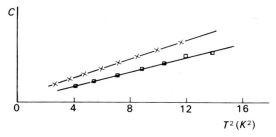

Fig. 10.11 The heat capacity of an insulator in two dimensions (a helium film) is proportional to the square of the temperature at sufficiently low temperature. The upper line is for a film of average thickness (1.13 monolayers) and the lower for 0.90 monolayers. After Brewer, 1970.

systems (Fig. 10.11). The T^3 law is of course consistent with the third law of thermodynamics, as was discussed in Section 4.4.

Debye introduced a most useful approximate theory of the lattice heat capacity which reduces correctly to the Dulong and Petit law in the high temperature limit and to the T^3 behaviour at low temperature and is usually a reasonable fit to the measurements at intermediate temperature.

The Debye approximation involves using eqn 10.52 for the dispersion law and therefore eqn 10.54 for the density of states function. These equations are however only correct in the long wavelength limit and the Debye density of states is quite different from the true density of states function of a real solid at higher frequencies as may be seen in Fig. 10.12. The approximation remains useful because only the integral of the density of states function is required for the lattice heat capacity.

The density of states function for an elastic continuum given by eqn 10.54 leads to an infinite number of normal modes if used in eqn 10.48. Debye

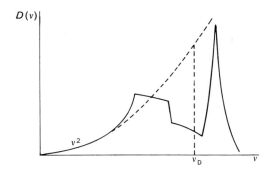

Fig. 10.12 The density of states function in the Debye approximation (dashed) and the actual density of states, for a typical solid.

introduced the discrete nature of the lattice by cutting off the density of states function at a characteristic frequency ν_D such that

$$\frac{12\pi V}{c_s^3} \int_0^{\nu_D} \nu^2 \, d\nu = 3N.$$

Therefore

$$\nu_D = \left(\frac{3Nc_s^3}{4\pi V}\right)^{\frac{1}{3}} \tag{10.56}$$

and the Debye temperature is given by

$$k\theta_D = h\nu_D. \tag{10.57}$$

The Debye temperature for a given solid can be found directly from the above equation (using an appropriate average value for the velocity of sound in the crystal) or by finding the value which gives the best agreement with experiment. The values of the Debye temperature found by these two methods usually agree to within 10 per cent, showing that the Debye theory has a certain validity. An exact agreement between the Debye theory and experimental heat capacities however is only obtained if the Debye temperature is allowed to change with temperature.

The energy of the lattice in the Debye approximation may be written (from eqns 10.47 and 10.54)

$$\bar{E} = \frac{12\pi V}{c_s^3} \int_0^{\theta_D} \frac{h\nu^3 \, d\nu}{e^{\beta h\nu} - 1}$$

$$= \frac{12\pi V}{c_s^3 h^3} (kT)^4 \int_0^{x_D} \frac{x^3 \, dx}{e^x - 1}$$

where

$$x_D = \frac{h\nu_D}{kT}.$$

The equation for the energy may be written in terms of the Debye temperature

$$\bar{E} = 9NkT \left(\frac{T}{\theta_D}\right)^3 \int_0^{x_D} \frac{x^3 \, dx}{e^x - 1}. \tag{10.58}$$

Now in the low-temperature limit ($T \ll \theta_D$) x_D goes to infinity and the integral becomes a definite integral with value $\pi^4/15$. The energy is then

$$\bar{E} = \frac{3\pi^4}{5} NkT \left(\frac{T}{\theta_D}\right)^3 \quad (T \ll \theta_D)$$

and the heat capacity

$$C_V = \frac{12\pi^4}{5} Nk \left(\frac{T}{\theta_D}\right)^3 \quad (T \ll \theta_D). \tag{10.59}$$

The slope of a graph of the measured heat capacity of an insulator as a function of the cube of the temperature therefore gives a value for θ_D. It is found that a true T^3 law only holds at temperatures below about $\theta_D/50$, which for most materials means in the liquid-helium range of temperature.

The Debye theory reduces to the law of Dulong and Petit in the extreme high-temperature $(T \gg \theta_D)$ limit. The exponential term in the integral of eqn 10.58 may be written as simply $(1+x)$ since x is now small and then

$$\bar{E} = 9NkT \left(\frac{T}{\theta_D}\right)^3 \int_0^{x_D} x^2 \, dx$$

$$= 3NkT \qquad (T \gg \theta_D)$$

as required. As in the Einstein model however it is clear that solids with a large θ_D will not reach the classical limit by room temperature. For example the Debye temperature for diamond is 2050 K (compared with the Einstein temperature of 1320 K) and the high-temperature limit is not attained even at the melting point ($\approx$ 3800 K).

The Debye theory does not provide exact agreement with the measured heat capacities of solids in the temperature region between $T > \theta_D/50$ and the high-temperature limit. The integral in eqn 10.58 has been tabulated and the difference between the Debye theory and experiment is usually expressed by calculating the value of θ_D at each temperature to give exact agreement with experiment. The Debye temperature would of course be a constant if the theory was correct but usually first decreases from the value extrapolated to absolute zero and then increases, becoming roughly constant at high temperature. The variation in θ_D is however usually less than plus or minus 20 per cent of the value at absolute zero so a 'typical' value provides reasonable agreement with experiment.

The Debye approximation to the density of states has been used to study many other phenomena in solid state physics such as compressibility, thermal expansion and elastic constants. The values of θ_D for a given system found by these different methods are usually in reasonable but not exact agreement.

The Debye approximation to the density of states function is so simple that it is commonly used in solid state calculations over the whole temperature range when great accuracy is not required. The prediction from eqn 10.58 that the lattice heat capacities of all solids lie on a universal curve when plotted as a function of T/θ_D is in remarkably good agreement with experiment (Fig. 10.13) although it has been seen that there is no unique value to take for θ_D for a material.

A better theory of the heat capacity cannot avoid the labour of actually evaluating the true density of states of the normal modes of the lattice which, as may be seen from Fig. 10.12, is quite different from the Debye approximation at high frequencies. It is also necessary to go beyond the approximation of simple harmonic oscillators to consider the anharmonic terms in the potential energy if exact agreement with experiment is to be obtained.

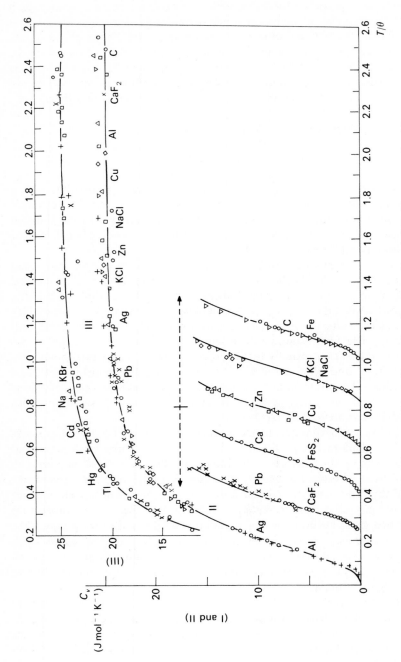

Fig. 10.13 Heat capacities of several substances $(J\,mol^{-1}\,K^{-1})$ compared with Debye's theory. For the sake of clarity, portions I and III are shown shifted. After Gopal, 1966.

The contribution of the free (conduction) electrons of metals to the heat capacity was briefly discussed in Section 10.1. The calculated heat capacity may be written

$$C_{el} = \frac{\pi^2}{2} Nk \frac{T}{T_F} = \gamma T$$

(eqn 10.18). At room temperature this contribution to the heat capacity was seen to be negligible relative to the lattice term with value $3Nk$. In the low temperature region however the lattice contribution decreases as T^3 and the coefficient γ can be found by experiment from the intercept on the C/T axis when plotted as a function of T^2.

$$\frac{C}{T} = AT^2 + \gamma$$

as shown for example in Fig. 4.2. The measured values of γ provide a direct comparison with the predictions of the quantum free-electron theory of metals since there are no unknown quantities in eqn 10.18 if each of the atoms in the crystal is assumed to contribute its valence electrons to the free-electron gas.

The agreement between the free-electron theory and experiment is normally expressed by defining an effective mass for the electrons (m^*) by the equation

$$\frac{m^*}{m} = \frac{\gamma_{ex}}{\gamma_{th}}$$

where γ_{th} is given by eqn 10.18 and γ_{ex} is found from the measured intercept shown in Fig. 4.2. The quantity m^* is found to be about 1.2 m for sodium and 1.3 m for copper showing that the conduction electrons cannot be treated as completely free electrons. The effective mass found from heat capacity measurements is in fact not a very useful quantity since the effective mass of an electron is different in general in different directions in the crystal and m^* is therefore some kind of average value.

The electronic contribution to the heat capacity therefore provides only limited information about the effective mass of an electron in a crystal lattice (or equivalently about the density of electron states at the Fermi energy) just as the lattice heat capacity was rather insensitive to the details of the density of states of the normal modes of lattice vibration. Heat capacity measurements are however relatively straightforward and even the rather limited information contained in θ_D or γ_{ex} is often useful when studying new materials.

The lattice contribution to the heat capacity is important for all solids and the electron contribution is important for all metals (and sometimes for semiconductors) but there are many other possible contributions to the heat capacities of particular solids. One other process common to all solids is the heat capacity due to the formation of defects in the perfect crystal lattice as the temperature is increased from absolute zero. The number of simple defects (n) in a crystal containing N atoms at temperature T was shown in Exercise 6.6 to be

$$n = N e^{-\beta \varepsilon} \quad (n \ll N)$$

where ε is the energy needed to form a defect. The defect energy is $n\varepsilon$ and the heat capacity

$$C = \frac{N\varepsilon^2}{kT^2} e^{-\beta \varepsilon}. \tag{10.60}$$

In most solids the energy of defect formation is sufficiently large for the defect heat capacity to be small right up to the melting point but in the solids formed by the so-called inert gases the heat capacity at the melting point is 20 per cent greater than the Dulong and Petit value for the lattice heat capacity and the excess heat capacity is well described by eqn 10.60.

The defect heat capacity is typical of many possible contributions to the heat capacity of solids which depend upon the increase in the disorder of the system as the temperature is increased from absolute zero. A binary alloy of two elements AB for example will (if in true thermodynamic equilibrium) form an ordered arrangement ABABAB in the crystal at absolute zero, as required by the third law of thermodynamics. At temperatures far above some characteristic temperature T_c the arrangement of the two types of atom becomes random. The entropy ($k \ln W$) of the alloy is zero in the ordered arrangement ($W = 1$) and equal to $Nk \ln 2$ (where N is the total number of atoms in the alloy) in the high temperature limit as in the case of the two-level system considered in Section 7.4.

The order–disorder process makes no contribution to the heat capacity at either high or low temperature since $T(\partial S / \partial T)_V$ is then equal to zero, but leads to a peak in the heat capacity at the temperature T_c where the long-range order vanishes (Fig. 10.14).

The order–disorder phenomenon also occurs when the lattice contains atoms with spin s. At sufficiently low temperature the spins must form an ordered arrangement in space in agreement with the third law of thermodynamics and at sufficiently high temperature are distributed at random over the allowed energy levels. The change in the magnetic entropy is therefore

$$\Delta S = Nk \ln (2s + 1). \tag{10.61}$$

Now the entropy change due to the magnetic process may be found by measuring the heat capacity over a wide temperature range, subtracting the estimated heat capacity due to other terms (notably the lattice thermal capacity) and using the thermodynamic relation

$$\Delta S = \int_0^T C \, d(\ln T).$$

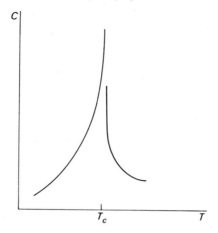

Fig. 10.14 The transition from an ordered binary alloy at low temperature to a disordered alloy leads to a peak in the heat capacity due to the increase in the entropy of the system.

The magnetic contribution (as in the order–disorder transformation shown in Fig. 10.14) is in fact largely concentrated near the transition temperature (Fig. 10.15) so the subtraction of the other terms is often straightforward. The spin of the particles can therefore be determined from measurements of the thermal capacity and eqn 10.61.

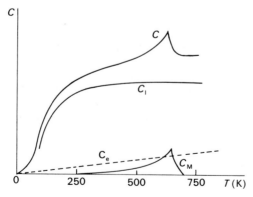

Fig. 10.15 A peak is observed in the heat capacity at the transition from ferromagnetism to paramagnetism (the Curie point). The measured heat capacity of nickel is shown and the estimated contributions of the lattice (C_l), electronic (C_e) and magnetic (C_M) contributions. Gopal, 1966.

One further contribution to the heat capacity of solids will be considered although this does not in any way exhaust the possibilities. A particle with spin s may be sufficiently weakly coupled to its neighbours for interparticle interactions to be negligible even at low experimental temperatures. The order–disorder phenomenon will not now be observed since the particles are only weakly interacting. One example of this type, the two-level system ($s = \frac{1}{2}$) was considered in Sections 6.3 and 7.2. A magnetic field splits the single allowed energy level into two levels separated by a distance Δ. At low temperature all the particles are in the lower level and at high temperature $N/2$ are in each level. The total entropy change is therefore $Nk \ln 2$ and the heat capacity shows a peak, as may be seen from eqn 7.45, repeated as eqn 10.62, and Fig. 10.16.

$$C = \frac{Nk\beta^2 \Delta^2 e^{\beta\Delta}}{(1 + e^{\beta\Delta})^2} \tag{10.62}$$

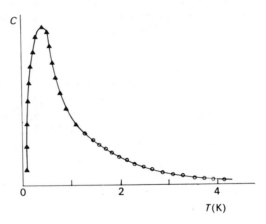

Fig. 10.16 The change in the populations of the states of a weakly interacting $(2I + 1)$-level system leads to a broad maximum in the heat capacity. The heat capacity of holmium is shown. At normal helium-temperatures it is only possible to observe the term in T^{-2} (eqn 10.63) but in this example the combination of a large magnetic field at the holmium nucleus and very low-temperature techniques allowed the whole curve to be measured. Communication from O. V. Lounasmaa.

where

$$\Delta = 2\mu\mu_0 H^*.$$

A peak of this type is called a Schottky anomaly. Notice that the peak is much broader than that of the order–disorder transition because the Schottky effect arises from each spin individually interacting with the magnetic field, but the order–disorder transition involves the interaction of all the particles in the system (as in the solid–liquid transition, Section 5.4).

The theory of the two-level Schottky effect ($s = 1/2$) can obviously be generalized to the case of spin s where the particles are distributed over ($2s + 1$) levels and comparison of theory and experiment provides a method for determining s. Alternatively if the magnetic moment (μ) is known for a nucleus or ion the magnetic field at the site may be found (as was noted in Section 7.4).

The theory of the Schottky effect at high temperature ($kT \gg \Delta$) leads to the result

$$C = \left(\frac{N\Delta^2}{2k} \right) \frac{1}{T^2} = DT^{-2} \tag{10.63}$$

(Exercise 7.2). The magnetic moment of a nucleus of spin I is so small that even at temperatures below 4 K the Schottky effect is usually only observed in the high-temperature limit. The total heat capacity would then be written

$$C = AT^3 + \gamma T + DT^{-2}$$

and the Schottky effect is observed as an increase in the heat capacity as the temperature decreases (Fig. 10.16).

The discussion of this section should have made clear both the importance and the limitations of the information which can be obtained from heat capacity measurements and in particular the importance of measurements in the liquid-helium temperature range below 4 K. In general heat capacity measurements are used to establish approximate values for such quantities as the effective mass of the conduction electrons in a metal or the spin and spacing of the energy levels of particles in a given solid. These values are then used in the design of other experiments (such as resonance techniques) which are capable of an accuracy many orders of magnitude greater than that possible in heat capacity measurements.

10.4 Phase transitions

The thermodynamics of a change of phase such as from liquid to vapour was considered in Section 5.4. Such a transition involves a latent heat, except at the critical point, and is called a first order phase transition because the first derivative of the Gibbs free energy per unit mass changes discontinuously at the transition, eqn 5.25. Along the coexistence curve for two phases, eqn 5.26,

$$\left(\frac{\partial P}{\partial T} \right)_\sigma = \frac{s_2 - s_1}{v_2 - v_1} = \frac{l}{T(v_2 - v_1)} \tag{10.64}$$

An unusual application of eqn 10.64, Clapeyron's equation, for the attainment of ultra-low temperature is described in Section 10.4.1.

At the critical point the distinction between liquid and vapour vanishes and the latent heat is zero at the phase transition. Such a transition is called a higher order phase transition, other examples are the ferromagnetic-para-magnetic transition (Fig. 10.15) and the lambda transition in liquid helium (Fig. 10.5). The statistical mechanics of higher order phase transitions has been one of the most active branches of theoretical physics in the last twenty years, culminating in the award of the Nobel prize to K. Wilson in 1982, but his theory (the renormalization group) is too difficult to present here. Instead the thermodynamics of the higher order phase transition is discussed in Section 10.4.2 and show, as usual, to provide the essential restraints on a successful detailed theory.

10.4.1 Pomeranchuk cooling

In 1950 Pomeranchuk proposed a new method for the attainment of very low temperature which has since been used (1973) to cool ^{3}He (Section 10.1) to 0.002 K, 2 mK, above absolute zero. The principle of the method may be understood using only Clapeyron's equation (10.64) and the P–T diagram for ^{3}He at temperatures below 300 mK (Fig. 10.17).

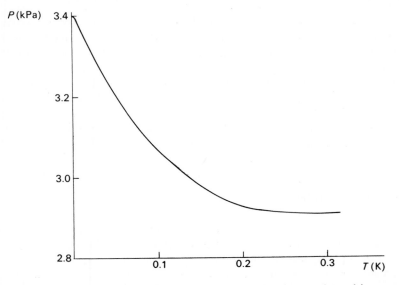

Fig. 10.17   The phase diagram of ^{3}He at low temperature, close to the melting curve. After Huiskamp and Lounasmaa, 1973.

The remarkable feature of Fig. 10.17 is that the P–T curve has a minimum. dP/dT is negative at temperatures below this minimum, but $(v_1 - v_s)$ remains positive. The entropy of the solid is therefore *greater* than that of the liquid (eqn 10.64) and an isentropic increase of pressure on the liquid leads to a decrease in the temperature of the system.

The P–T and S–T diagrams are shown for temperatures below 30 mK in Figs. 10.18 and 10.19. An isentropic increase of pressure from X to Y at a starting temperature of 25 mK would lead to a final temperature of 2 mK. At 2 mK all the liquid has been converted to solid and the cooling process is complete. It is clear from the S–T diagram that the starting temperature must be far below 318 mK, where the solid and liquid entropies are equal, if appreciable cooling is to be obtained. Temperatures in the 20–30 mK region can now be reached fairly easily however using ^{3}He–^{4}He refrigerators.

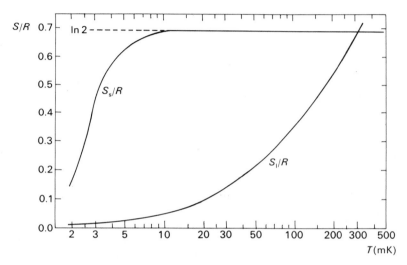

Fig. 10.18 The entropy of solid (S_s) and liquid (S_1) ^{3}He along the melting curve. The full nuclear spin entropy is marked in the figure. After Huiskamp and Lounasmaa, 1973.

The main features of Pomeranchuk cooling have been discussed so far using only thermodynamic arguments but statistical mechanics is required to explain the shape of the S–T curves. ^{3}He is a Fermi particle (Section 10.1) and the liquid is an example of a Fermi liquid with $T_F = 450$ mK. In Section 10.1 it was shown that the entropy of a Fermi gas is proportional to the temperature for $T \ll T_F$ since

$$S = \int \frac{C_V}{T} \, dT = \gamma \int dT = \gamma T \qquad (10.65)$$

and this result also appears to hold for a Fermi liquid (as shown in Fig. 10.19). The entropy of the solid however is largely due to the disorder of the nuclear spins of ^{3}He. Since $I = \frac{1}{2}$ for ^{3}He the 'high-temperature limit' of the entropy is $R \ln 2$ per mole (Section 6.3) and the entropy decreases exponentially to zero at low temperature. The liquid entropy reaches the value $R \ln 2$ at 318 mK,

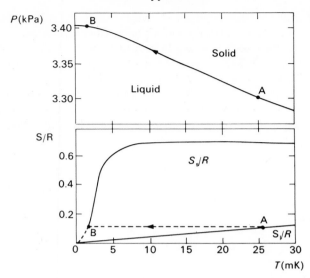

Fig. 10.19   The phase and entropy diagrams of ^{3}He demonstrating Pomeranchuk cooling. After Huiskamp and Lounasmaa, 1973.

the minimum in the coexistence curve, when dP/dT and $(s_s - s_l)$ are equal to zero.

   Pomeranchuk cooling is the obvious method to use when studying the properties of ^{3}He at low temperature since the liquid cools itself, but it is more difficult to cool other materials because solid ^{3}He forms in the *warmest* part of the cell and thermally isolates the sample. The other main experimental difficulty of the method is the necessity for pressure changes to be made reversibly if frictional heating is not to outweigh the Pomeranchuk cooling. Two good references to this important topic will be found in the reading list.

10.4.2 Critical point exponents for higher-order phase transitions

The density of liquid (ρ_1) and gas (ρ_g) along the coexistence curves shown in Fig. 10.20 are different below the critical point. The quantity ($\rho_1 - \rho_g$) which vanishes smoothly at the critical point is called the *order parameter* for the liquid–gas transition. In recent years it has come to be realised that many other higher-order transitions (for example ferromagnetic, superconducting, superfluid) can be characterised by an order parameter and that all these transitions have features in common, sufficiently close to the critical point. The analogy between the liquid–gas transition at the critical point and magnetic transitions is particularly close. The apparent simplicity of the behaviour of systems close to the critical point which will be discussed in this section, has led to much theoretical work in recent years. The statistical mechanics of the critical region is too difficult to discuss in this book but one derivation will be given of a most important thermodynamic result.

The modern study of critical phenomena may be said to have begun in 1945 when Guggenheim showed that measurements on simple gases below the critical temperature were well described by a law of corresponding states of the form

$$\frac{\rho_1 - \rho_g}{2\rho_c} = A\left(1 - \frac{T}{T_c}\right)^\beta \quad (T_{\to T_{c-}}) \tag{10.66}$$

$$= At^\beta$$

where A and β are constants and ρ_c is the critical density. The constant β is called a critical point exponent. The expression T_{c-} means that the temperature approaches the critical point from below. The law of corresponding states was seen in Section 2.2 to be predicted by any two-parameter equation of state such as the van der Waals equation but the experimental value of β is found to be close to 1/3 rather than 1/2 as predicted for a van der Waals gas.

The approximate equations of state discussed in Section 2.2 are therefore inadequate in the critical region and many attempts have been made to construct more realistic equations of state.

The isothermal compressibility has already been seen to diverge at the critical point since

$$\kappa_T = -\frac{1}{V}\left(\frac{\partial V}{\partial P}\right)_T = \frac{1}{\rho}\left(\frac{\partial \rho}{\partial P}\right)_T \tag{10.67}$$

and $(\partial P/\partial V)_T$ is zero at the critical point. The van der Waals equation predicts that above the critical temperature but on the critical isochore (Fig. 10.20)

$$\kappa_T \approx B|t|^{-\gamma} \quad (T_{\to T_{c+}}; \rho = \rho_c) \tag{10.68}$$

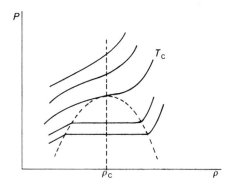

Fig. 10.20 The isotherms for gas and liquid as a function of pressure and density. The critical isochore is shown as a dotted (vertical) line.

with γ equal to unity. At temperatures below T_c the isothermal compressibility has to be defined for each phase separately and

$$\kappa_T \approx B't^{-\gamma'} \quad (T \to T_{c-}; \rho = \rho_L \text{ or } \rho_g) \tag{10.69}$$

along the coexistence curve. The van der Waals equation leads to $\gamma = \gamma' = 1$ but the experimental results suggest that γ' is about 1.2 for most gases and is not necessarily equal to γ. The standard notation of primed exponents (γ') for temperatures below T_c should be noted; β however is an exception to the rule.

The experimental specific heat of a system at constant volume along the critical isochore (Fig. 10.21) may be written

$$C_V = -A^{\pm} \ln |t| + B^{\pm} \tag{10.70}$$

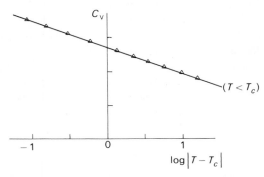

Fig. 10.21 The heat capacity at constant volume of argon along the critical isochore appears to go to infinity at the critical temperature.

where A and B are constants. The specific heat at constant volume therefore also appears to go (rather slowly) to infinity.

A more general form for eqn 10.70, which below T_c involves the critical point exponent α', follows from the definition

$$-\alpha' = \lim_{T \to T_{c-}} \left(\frac{\ln C_V}{\ln t} \right) \quad (\rho = \rho_c) \tag{10.71}$$

The eqn 10.71 reduces to eqn 10.70 for the special case of α' equal to zero. The experimental values for α' are close to 0.12 e.g. $CO_2 = 0.125$, $^4He = 0.127$. The best theoretical value is 0.11. The van der Waals prediction is $\alpha = \alpha' = 0$.

The other critical point exponents should also be understood as arising from limits defined in the manner of eqn 10.71. For example the equation (10.68) for the isothermal compressibility is not to be understood as exact (hence the $\approx$ sign) but as indicating that sufficiently close to T_c the divergence of the term in γ is stronger than that of any other term.

The decision about the size of the region which is 'sufficiently close' to T_c is often difficult to make when analysing experimental results. At very small values of $(T_c - T)/T_c$ the choice of T_c may have an appreciable effect on the value found for the critical point exponent and at larger values the measurements, may no longer be within the critical region. A new approach to this problem is discussed in Exercise 10.15.

Accurate measurements are difficult to perform in the critical region since the liquid–gas transition is found to be extremely sensitive to the effects of gravity (due to the large value of the isothermal compressibility) and the time for the system to come to thermodynamic equilibrium also becomes very long as the critical point is approached. The infinities associated with the critical point exponents, such as eqn 10.68, are not of course observed by experiment, partly because no real system has a uniquely defined value for T_c, due to impurities (although in favourable cases such as superconductivity the transition may take place within a range of 10^{-3} K) but also for the more fundamental reason that eqn 10.68 only applies in the thermodynamic limit. In a finite system the transition will always be 'rounded' close to T_c due to the effect of fluctuations.

The calculation of the critical point exponents using the methods of statistical mechanics is a difficult branch of current research and will not be considered here. A number of inequalities obeyed by the various critical point exponents can however be derived using only thermodynamics and as an example the inequality

$$\alpha' + 2\beta + \gamma' \geqslant 2 \tag{10.72}$$

will be considered. These inequalities are obviously of great importance since they represent a test of the properties of any model of the phase transition used by statistical mechanics and are also a useful check on the experimental values of the critical point exponents. It appears possible in fact that equation 10.72 is an equality but no rigorous proof has yet been given. The original proof of the inequality was given by Rushbrooke (1963) for the analogous magnetic transition and was extended to the liquid–gas transition by Fisher (1964).

The proof of the inequality involves the construction of an exact equation for the specific heat at constant volume of a system along the critical isochore. All the terms of this equation are then recognised to be positive and by excluding all but one of these terms the inequality is established. The proof is fairly long, but straightforward, and is given in some detail since it is a good example of the power of thermodynamics.

The system under consideration has total mass M contained in a constant volume V. Since the system is to be heated along the critical isochore, Fig. 10.20, the relationship between M and V is determined by the equation

$$Mv_c = V \tag{10.73}$$

where v_c is the volume of unit mass of substance at the critical point (the reciprocal of the critical density). At any temperature below T_c the system will have a mass m_1 in the liquid phase with volume per unit mass v_1 and mass m_g in the gas phase with volume per unit mass v_g where

$$m_1 + m_g = M$$
$$m_1 v_1 + m_g v_g = V = M v_c. \qquad (10.74)$$

The mole fractions in the two phases may therefore be written

$$x_1 = \frac{m_1}{m_1 + m_g} = \frac{v_g - v_c}{v_g - v_1} \qquad (10.75)$$

$$x_g = \frac{m_g}{m_1 + m_g} = \frac{v_c - v_1}{v_g - v_1}. \qquad (10.76)$$

The effect of an increase in temperature δT is to transfer a mass δm from the liquid to the gas phase. The specific heat of the liquid phase will be written C_σ^1 where the subscript σ means along the coexistence curve. Then the total specific heat at constant volume of the whole system (C_V) will be given by the equation

$$(m_1 + m_g) C_V \delta T = (m_1 C_\sigma^1 + m_g C_\sigma^g) \delta T + l \delta m \qquad (10.77)$$

where l is the latent heat.

The quantity δm may be found from the condition that the total volume remains constant using eqn 10.74

$$m_1 v_1 + m_g v_g = (m_1 - \delta m)\left[v_1 + v_1 \left(\frac{\partial v_1}{\partial T}\right)_\sigma \delta T\right] + (m_g + \delta m)\left[v_g + v_g \left(\frac{\partial v_g}{\partial T}\right)_\sigma \delta T\right].$$

Therefore

$$\delta m = \frac{m_1 \left(\frac{\partial v_1}{\partial T}\right)_\sigma + m_g \left(\frac{\partial v_g}{\partial T}\right)_\sigma}{v_1 - v_g} \delta T. \qquad (10.78)$$

On substituting for δm using eqn 10.78 for l using the Clapeyron equation 10.64 and for x_1 and x_g, eqn 10.77 becomes

$$C_V = x_1 C_\sigma^1 + x_g C_\sigma^g - T\left(\frac{\partial P}{\partial T}\right)_\sigma \left[x_1 \left(\frac{\partial v_1}{\partial T}\right)_\sigma + x_g \left(\frac{\partial v_g}{\partial T}\right)_\sigma\right]. \qquad (10.79)$$

The quantities C_σ^1 and C_σ^g now have to be eliminated from eqn 10.79. In one phase the standard thermodynamic relation

$$T dS = T\left(\frac{\partial S}{\partial T}\right)_V dT + T\left(\frac{\partial S}{\partial V}\right)_T dV$$

leads to

$$T\left(\frac{\partial S}{\partial T}\right)_\sigma = T\left(\frac{\partial S}{\partial T}\right)_V + T\left(\frac{\partial S}{\partial V}\right)_T \left(\frac{\partial V}{\partial T}\right)_\sigma.$$

Therefore

$$C_\sigma = C_V + T\left(\frac{\partial P}{\partial T}\right)_V \left(\frac{\partial V}{\partial T}\right)_\sigma$$

where the Maxwell relation has been used to eliminate $(\partial S/\partial V)_T$.

The term $(\partial P/\partial T)_V$ may be transformed by the equation

$$\left(\frac{\partial P}{\partial T}\right)_\sigma = \left(\frac{\partial P}{\partial T}\right)_V + \left(\frac{\partial P}{\partial V}\right)_T \left(\frac{\partial V}{\partial T}\right)_\sigma$$

to give

$$C_\sigma = C_V - T\left(\frac{\partial P}{\partial V}\right)_T \left(\frac{\partial V}{\partial T}\right)_\sigma^2 + T\left(\frac{\partial P}{\partial T}\right)_\sigma \left(\frac{\partial V}{\partial T}\right)_\sigma. \tag{10.80}$$

This equation applies to each phase separately so on reintroducing the superscripts l and g and substituting into eqn 10.79 it is found that the second term in eqn 10.80 cancels leaving

$$C_V = x_1 C_V^l + x_g C_V^g - T\left[x_1\left(\frac{\partial P}{\partial v_1}\right)_T \left(\frac{\partial v_1}{\partial T}\right)_\sigma^2 + x_g\left(\frac{\partial P}{\partial v_g}\right)_T \left(\frac{\partial v_g}{\partial T}\right)_\sigma^2 \right].$$

It is now convenient to introduce the isothermal compressibility of each phase by the usual definition (eqn 10.67) and to substitute the density of each phase for the reciprocal of the volume per unit mass. Therefore

$$C_V = x_1 C_V^l + x_g C_V^g + \frac{x_1 T}{\rho_1^3 \kappa_T^l}\left(\frac{\partial \rho_1}{\partial T}\right)_\sigma^2 + \frac{x_g T}{\rho_g^3 \kappa_T^g}\left(\frac{\partial \rho_g}{\partial T}\right)_\sigma^2. \tag{10.81}$$

This is the final (exact) expression for the specific heat of the whole system at constant volume along the critical isochore. The reason for introducing C_V^l and C_V^g is that specific heats at constant volume are always positive since (eqn 7.18)

$$C_V = \frac{\overline{[(E - \bar{E})]^2}}{kT}$$

and all the terms on the right-hand side are positive. The third term on the right-hand side of eqn 10.81 is also positive so the inequality

$$C_V \geqslant \frac{x_g T}{\rho_g^3 \kappa_T^g}\left(\frac{\partial \rho_g}{\partial T}\right)_\sigma^2 \tag{10.82}$$

has been established. As the critical point is approached at the constant total density (ρ_c) the value of x_g approaches $1/2$, $(v_c = (v_1 + v_g)/2)$ and ρ_g approaches ρ_c. The isothermal compressibility and $(\partial \rho_g/\partial T)_\sigma$ however, diverge near the critical point since

$$\left(\frac{\partial \rho_g}{\partial T}\right) \approx t^{\beta - 1}$$

and hence

$$\ln C_V \geqslant (\gamma' + 2\beta - 2)\ln t$$
$$\ln C_V \geqslant (2 - 2\beta - \gamma')|\ln t| + \ldots (T_{\to T_{c-}}).$$

since $\ln t < 0$.
Therefore

$$\frac{\ln C_V}{|\ln t|} \geqslant 2 - 2\beta - \gamma'. \tag{10.83}$$

The higher-order terms vanish in the limit and the left-hand side is then the definition of α' so

$$\alpha' + 2\beta + \gamma' \geqslant 2. \tag{10.83}$$

There have been a number of attempts to show that on the basis of further assumptions (the Scaling laws) the equality is in fact correct. The van der Waals equation satisfies the equality ($\alpha' = 0$, $\beta = 1/2$, $\gamma' = 1$) but the agreement with experiment for β is so poor that this result has little significance. The values of the critical exponents for a number of idealised theoretical models have however been calculated with great accuracy and do seem to support the equality. The equality is also compatible with experiment although this is not a rigorous test since γ' cannot be measured to high accuracy.

There are many other thermodynamic inequalities but only one further example will be given here. The Griffiths inequality (1965) may be written

$$\alpha' + \beta(1 + \delta) \geqslant 2 \tag{10.84}$$

where α' and β have already been defined and δ is defined by the variation of $(P - P_c)$ with $(\rho - \rho_c)$ along the critical isotherm T_c. The early measurements on ^{4}He in the critical region were found to be in disagreement with this inequality ($\alpha' = 0.017$, $\beta = 0.354$, $3.8 \leqslant \delta \leqslant 4.1$) but further work suggests that $\alpha' \approx 0.13$ leading to agreement, within experimental error, with eqn 10.84. The magnitude of the change in the experimental value of α' should be sufficient evidence of the difficulty of measurements near the critical point and of the importance of the thermodynamic inequalities as a check on such measurements. A full account of the theory of the critical region may be found in Reichl (1980).

10.5 Magnetism

The study of the magnetic and dielectric properties of materials is one of the most important applications of the methods of thermodynamics and statistical mechanics. Magnetism is inherently a quantum phenomenon however and a full description of the magnetic properties of solids requires a greater knowledge of quantum mechanics and solid state physics than has been assumed in the rest of this book. Only the simplest statistical model will therefore be treated in this section, although the model will be seen in Section

10.5.5 to be sufficient to explain the principle of two important applications in low-temperature physics. The thermodynamics of magnetism is also a difficult topic and only an outline of the problem is given in Section 10.5.3.

10.5.1 Simple paramagnetism of localized spins

The simplest example of a paramagnetic substance is an array of N_s localized particles each with spin $\frac{1}{2}$ in a volume V. The interaction between the spins and between a given spin and the crystal lattice must be sufficient to allow the system to come to thermal equilibrium but has no other effect on the individual spins. In the next section it will be seen how this description has to be modified when dipolar forces between the spins are important.

The spin-$\frac{1}{2}$ system has already been briefly discussed in Section 7.4. The single-particle energy level is two-fold degenerate in zero external magnetic field and splits into two non-degenerate levels with energy $\mu\mu_0 H^*$ and $-\mu\mu_0 H^*$ when the sample is placed in a uniform field with value $\mu_0 H^*$ where μ is the dipole moment of the particle. It should be immediately apparent that this description is inadequate for a real solid as $T_{\to 0\,\mathrm{K}}$ since the entropy of the system in zero external field would then be $N_s k \ln 2$ but if the spins formed an *ordered* arrangement in space the entropy would go to zero. When the spins in a region of the sample are all aligned parallel to each other, for example, the system is said to be ferromagnetic.

The interactions between the spins cannot therefore be ignored except at sufficiently high temperature where the interaction energy between two dipoles a distance a apart is much less than the thermal energy

$$\frac{\mu^2 \mu_0}{4\pi a^3} \ll kT. \tag{10.86}$$

This condition is obviously best satisfied for particles of small dipole moment which are widely separated in space.

The dipole moment of a nucleus is of order

$$\mu_N = \frac{eh}{4\pi m_p} = 5.0_5 \times 10^{-27} \, \mathrm{J\,T^{-1}}$$

where m_p is the mass of the proton. (The values for particular isotopes vary between $5\mu_N$ and zero but these distinctions will not be considered here.) The two sides of eqn 10.90 are equal at a temperature of 7×10^{-9} K for a system with a moment of μ_N on each site, that is a spacing of about 0.3 nm. A system in which the magnetism is entirely due to the nuclei (a nuclear paramagnet) therefore satisfies the inequality under all normal circumstances.

The dipole moment of an atom or ion with spin $\frac{1}{2}$ is of order

$$\mu_B = \frac{eh}{4\pi m_e}$$

where m_e is the mass of the electron and the inequality is now satisfied for temperatures greater than 0.02 K if a spin is placed on each lattice site. This result is however seriously misleading because when the wave functions of adjacent atoms *overlap* an interaction of quantum mechanical origin called the exchange interaction becomes far more important than the dipolar interaction. In iron for example the exchange interaction is sufficiently strong for the system to be ferromagnetic at temperatures below 10^3 K.

A low-temperature electronic paramagnet must therefore be formed by *diluting* the N_s magnetic ions with N non-magnetic ions. A famous example of such a crystal which will be discussed in Section 10.6.4 is cerium magnesium nitrate (CMN) which has the formula $2\,Ce(NO_3)_3 . \, 3\,Mg(NO_3)_2 . \, 24\,H_2O$. Only the cerium ions have a magnetic moment so the dilution is sufficient to prevent the formation of an ordered system down to temperatures below 10 mK.

The magnetic properties of the simple paramagnet can now quickly be deduced but to stress the specialised nature of the results, the subscript ni for non-interacting spins will be used throughout this section.

The total magnetic moment of the sample in a uniform external field, $\mu_0 H^*$ is simply equal to the product of the moment of one particle and the difference in the number of spins in each level

$$\mathscr{M}_{ni} = N_s \mu \left(\frac{1 - e^{-\beta\Delta}}{1 + e^{-\beta\Delta}} \right) \tag{10.87}$$

$$\Delta = 2\mu\mu_0 H^*.$$

The magnetization is defined as the magnetic moment per unit *volume* of sample

$$M_{ni} = n_s \mu \left(\frac{1 - e^{-\beta\Delta}}{1 + e^{-\beta\Delta}} \right) \tag{10.88}$$

and the isothermal susceptibility is defined by the equation

$$\chi_T = \left(\frac{\partial M}{\partial H^*} \right)_T . \tag{10.89}$$

The form of equation 10.87 is shown in Fig. 10.22. In the high-temperature limit

$$M_{ni} = \frac{n_s \mu^2 \mu_0 H^*}{kT} = f\left(\frac{H^*}{T} \right) \tag{10.90}$$

$$\chi_{ni} = \frac{n_s \mu^2 \mu_0}{kT} = \frac{C}{T} \tag{10.91}$$

a result known as Curie's Law. The value of C is $\approx 0.3(n_s/n)$K for an electronic paramagnet and $9 \times 10^{-8}(n_s/n)$K for a nuclear paramagnet so the

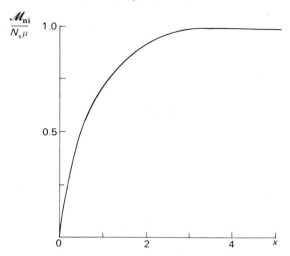

Fig. 10.22 The magnetization of a simple paramagnet as a function of $x = \mu\mu_0 H^*/kT$.

magnetic susceptibility is small at normal temperature. A system which obeys equation 10.91 is said to be an *ideal paramagnet*.

The susceptibility is independent of the external field in the high-temperature region where Curie's Law applies but this is not generally true. The next term in the high-temperature expansion of eqn 10.88 for example leads to

$$M_{ni} = \frac{n_s\mu^2\mu_0 H^*}{kT} - \frac{n_s\mu^4\mu_0^3 H^{*3}}{3k^3 T^3} \tag{10.92}$$

$$\chi_{ni} = \frac{n_s\mu^2\mu_0}{kT} - \frac{n_s\mu^4\mu_0^3 H^{*2}}{k^3 T^3}. \tag{10.93}$$

The susceptibility is now a function of both temperature and field but from considerations of symmetry it is clear that the susceptibility will always be an even function of the external field and the magnetisation an odd function.

A comparison of the partition function for the non-interacting spin system of N_s spins each with spin $\frac{1}{2}$

$$Z_s = \left[e^{\frac{-\beta\Delta}{2}} + e^{\frac{\beta\Delta}{2}} \right]^{N_s}$$

with eqn 10.87 leads to the result

$$\mathcal{M}_{ni} = \frac{kT}{\mu_0} \left(\frac{\partial \ln Z_s}{\partial H^*} \right)_T \tag{10.94}$$

which may be used to extend the results for spin $\frac{1}{2}$ to particles of higher spin.

The application of the equations developed in this section will be found in Sections 10.5.4 and 10.5.5 but in the next section their range of usefulness for real magnetic systems will be discussed.

10.5.2. Dipole–dipole interactions

In the previous section each spin was assumed to interact directly with the external field $\mu_0 H^*$. This approximation is satisfactory if the magnetization is very small, as for a nuclear paramagnet, but in this section the corrections needed when dipole–dipole interactions are non-negligible will be considered.

When an ellipsoid is placed in a uniform magnetic field $\mu_0 H^*$ and a magnetization M is produced, the *macroscopic* internal field is given by

$$H_i = H^* - \gamma_{\mathrm{d}} M \tag{10.95}$$

where γ_{d} is the demagnetizing factor of the sample in the direction of the field. The demagnetizing factors along the principle axes of the ellipsoid are subject to the sum rule

$$\gamma_{\mathrm{d}}^x + \gamma_{\mathrm{d}}^y + \gamma_{\mathrm{d}}^z = 1 \tag{10.96}$$

so γ_{d} is equal to $\frac{1}{3}$ for a sphere, by considerations of symmetry, to zero along a long rod and to $\frac{1}{2}$ transverse to a long rod of circular cross-section.

The macroscopic internal field is the field used in, for example the Maxwell electromagnetic equations but is not correct for the calculation of the behaviour of *localized* spins. The dipoles occupy a special position in the unit cell and the *local* field at each lattice site must be found by adding to eqn 10.96 a contribution due to all the other dipoles in the lattice. An elementary account of the local field will be found in Kittel (1986) and a fuller analysis in Robinson (1973). The simplest correction term is due to Lorentz and adds to eqn 10.96 a term $\gamma_{\mathrm{L}} M$ where γ_{L} is equal to $\frac{1}{3}$. Therefore

$$H_i^{\mathrm{loc}} = H^* + (\gamma_{\mathrm{L}} - \gamma_{\mathrm{d}}) M \tag{10.97}$$

and since all three vectors lie along the same direction in the systems under consideration they may be written as scalars.

The magnetization of the system of spins interacting via dipolar forces in the external field $\mu_0 H^*$ is equal to that of the non-interacting system in the local field $\mu_0 H_i^{\mathrm{loc}}$

$$M_{\mathrm{int}}(H^*) = M_{\mathrm{ni}}(H_i^{\mathrm{loc}}) = M_{\mathrm{ni}}(H^* + (\gamma_{\mathrm{L}} - \gamma_{\mathrm{d}}) M_{\mathrm{int}}).$$

The right-hand side may be expanded in a Taylor series, the second term of which is zero, since M must be an odd function of H^*, and higher terms are negligible except for strongly magnetic substances so

$$M_{\mathrm{int}}(H^*) = M_{\mathrm{ni}}(H^*) + (\gamma_{\mathrm{L}} - \gamma_{\mathrm{d}}) M_{\mathrm{int}}(H^*) \left[\frac{\partial M_{\mathrm{ni}}(H^*)}{\partial H^*} \right]_T.$$

Therefore

$$M_{\text{int}}(H^*) = \frac{M_{\text{ni}}(H^*)}{1-(\gamma_{\text{L}}-\gamma_{\text{d}})\chi_{\text{ni}}}. \tag{10.98}$$

There are unfortunately at least *two* possible definitions of the susceptibility of a system of interacting spins and although these will be distinguished by superscripts in this section the reader should note that in most books no attempt is made to distinguish between them. The response of the magnetization given by eqn 10.98 to an external field is clearly shape-dependent so the susceptibility of unit volume of a given sample may be written

$$\chi_{\text{int}}^{\text{sample}} = \left(\frac{\partial M_{\text{int}}}{\partial H^*}\right)_T = \frac{\chi_{\text{ni}}}{1-(\gamma_{\text{L}}-\gamma_{\text{d}})\chi_{\text{ni}}}. \tag{10.99}$$

The sample susceptibility of a sphere is therefore equal to that of the non-interacting spins and in the region in which Curie's law applies

$$\chi_{\text{int}}^{\text{sphere}} = \frac{C}{T} \tag{10.100}$$

an important result which will be discussed in Section 10.6.4.

The susceptibility of the *material*, independent of the shape of the sample however is given by

$$\chi_{\text{int}}^{\text{material}} = \left(\frac{\partial M_{\text{int}}}{\partial H_i}\right)_T = \frac{\chi_{\text{ni}}}{1-\gamma_{\text{L}}\chi_{\text{ni}}} = \frac{C}{T-\gamma_{\text{L}}C} \tag{10.101}$$

where H_i is given by eqn 10.95.

The value of χ_{ni} is always so small for nuclear paramagnets under normal experimental conditions that eqns 10.101 and 10.99 are equivalent to eqn 10.91. However an ingenious experiment by Abragam and coworkers has recently shown that if the spin temperature (Section 7.8) is reduced to about $1\,\mu\text{K}$ then the nuclear spins do form a ferromagnetic (or other ordered) system.

As has already been remarked the most important interaction between electron spins is not normally the dipolar coupling, so although eqn 10.101 is of the form of the experimental Curie–Weiss law

$$\chi = \frac{C}{T-T_p} \tag{10.102}$$

where T_p is a constant for a given material, the paramagnetic Curie temperature, the value of T_p is usually much larger than the value of $0.1\,(n_{\text{s}}/n)\text{K}$ calculated from eqn 10.101. Only for the most dilute electronic paramagnets such as CMN is eqn 10.101 useful.

10.5.3 *Thermodynamics of magnetic systems*

The second law of thermodynamics was written in Chapter 3 as

$$T dS = dU + \sum_r Y_r dy_r$$

where Y_r is an intensive variable for example pressure or tension and y_r the conjugate extensive variable (volume, length) such that $Y_r dy_r$ is a work term. In many cases only one work term was found to be of importance and the equation

$$T dS = dU + Y dy$$

was sufficient to describe a simple system in which the state of the system was defined by just two independent parameters.

The application of a magnetic field to a solid usually leads to only a small change in the energy of the system relative to the lattice energy so the volume change due to the magnetization can be neglected. (This is not true near the Curie point of a ferromagnetic material but this point will not be considered here.) The magnetization of a paramagnetic (or diamagnetic) substance is a single valued function of the external magnetic field so the system is a simple system as defined in Chapter 2.

Fig. 10.23 A possible system for performing reversible work on a sample in the field of a permanent magnet. The system of interest consists only of the sample.

A number of different expressions may be written for the magnet work term, depending upon the definition of the system of interest. Only the simplest, and most important, case will be considered here; a fuller discussion is given in Kittel and Kroemer (1980) or Robinson (1973).

The force of attraction between a small ellipsoid with induced magnetic moment $\mathcal{M}(x)$ at a point x and a permanent magnet may be written

$$F = \mu_0 \,\mathcal{M}(x).\frac{\partial H^*(x)}{\partial x}$$

where $\mu_0 H^*(x)$ is the field due to the magnet in the absence of the sample.

One possible method of bringing the sample from infinity to a point x_0 by a quasi-static process is shown in Fig. 10.23. The work done on the sample is

$$W_A = -\mu_0 \int_\infty^{x_0} \mathcal{M}(x) \cdot \frac{\partial H^*(x)}{\partial x} dx = -\mu_0 \int_0^{\mu_0 H^*(x_0)} \mathcal{M}(x) dH^*.$$

Therefore

$$dW_A = -\mu_0 \mathcal{M} \cdot dH^* \tag{10.103}$$

and the Helmholtz free energy of the magnetic part of the system may be written

$$\begin{aligned} dF_A &= -SdT + dW \\ &= -SdT - \mu_0 \mathcal{M} dH^* \end{aligned} \tag{10.104}$$

where the component of $\mathcal{M}$ parallel to H^* is to be used in eqn 10.104. It is shown in Kittel and Kroemer (1980) and Robinson (1973) that this form for the Helmholtz free energy is related to the partition function by the usual expression

$$F_A = -kT \ln Z.$$

Therefore

$$\mathcal{M} = -\frac{1}{\mu_0} \left(\frac{\partial F_A}{\partial H^*} \right)_T = \frac{kT}{\mu_0} \left(\frac{\partial \ln Z}{\partial H^*} \right)_T \tag{10.105}$$

in agreement with the special result already obtained for a simple paramagnetic substance. The effect of interactions between the spins must therefore be included in the Hamiltonian of the system and hence in the partition function.

The magnetic analogues of the Maxwell relations derived in Section 4.2 may be established from eqn 10.104. The most useful are

$$\left(\frac{\partial S}{\partial H^*} \right)_T = \mu_0 \left(\frac{\partial \mathcal{M}}{\partial T} \right)_{H^*} \tag{10.106}$$

derived directly from eqn 10.104 and since

$$\left(\frac{\partial S}{\partial H^*} \right)_T = \left(\frac{\partial S}{\partial \mathcal{M}} \right)_T \left(\frac{\partial \mathcal{M}}{\partial H^*} \right)_T$$

$$\left(\frac{\partial S}{\partial \mathcal{M}} \right)_T = -\mu_0 \left(\frac{\partial H^*}{\partial T} \right)_{\mathcal{M}}. \tag{10.107}$$

The application of these two Maxwell relations to the ideal paramagnet with the equation of state

$$\mathcal{M} = \frac{VCH^*}{T}$$

will now be considered.

According to the third law of thermodynamics, the left-hand side of eqn 10.106 must go to zero as $T_{\to 0K}$, so an ideal paramagnet cannot exist in this limit, in agreement with the special result obtained in eqn 10.87. Secondly, the change in the magnetic entropy of an ideal paramagnet is a function only of the magnetization since

$$S_f - S_i = -\mu_0 \int \left(\frac{\partial H^*}{\partial T} \right)_{\mathcal{M}} d\mathcal{M} = \frac{-\mu_0}{2CV} (\mathcal{M}_f^2 - \mathcal{M}_i^2). \qquad (10.108)$$

Two final results, valid for any paramagnet, will be derived in this section. Firstly, a reversible increase in the magnetization of a paramagnetic material at constant temperature leads to a flow of heat out of the material. Considering the entropy as a function of the magnetization and the temperature

$$T dS = T \left(\frac{\partial S}{\partial T} \right)_{\mathcal{M}} dT + T \left(\frac{\partial S}{\partial \mathcal{M}} \right)_T d\mathcal{M}$$

so at constant temperature

$$T\Delta S = \Delta Q = T \left(\frac{\partial S}{\partial \mathcal{M}} \right)_T \Delta \mathcal{M} = -\mu_0 T \left(\frac{\partial H^*}{\partial T} \right)_{\mathcal{M}} \Delta \mathcal{M} \qquad (10.109)$$

$(\partial H^*/\partial T)_{\mathcal{M}}$ is always positive for a paramagnet so heat always flows out of the system when the magnetization is increased at constant temperature.

Secondly, an isentropic decrease of the external magnetic field always leads to a decrease in the temperature of a paramagnet. Considering the entropy as a function of temperature and magnetic field

$$\begin{aligned} T dS &= T \left(\frac{\partial S}{\partial T} \right)_{H^*} dT + T \left(\frac{\partial S}{\partial H^*} \right)_T dH^* \\ &= C_H dT + \mu_0 T \left(\frac{\partial \mathcal{M}}{\partial T} \right)_{H^*} dH^* \end{aligned}$$

where C_H is the heat capacity in constant external magnetic field. The temperature change due to a small isentropic change in the magnetic field is

$$\Delta T = \frac{-T \left(\frac{\partial \mathcal{M}}{\partial T} \right)_{H^*} \mu_0 \Delta H^*}{C_H} \qquad (10.110)$$

$(\partial \mathcal{M}/\partial T)_{H^*}$ is always negative at temperatures above absolute zero for a simple ferro- or paramagnetic material so the temperature of the sample will tend to decrease when the magnetic field is reduced.

These two results are important in the theory of adiabatic demagnetization which is discussed in Section 10.5.5.

10.5.4 The thermodynamic temperature scale at low temperature

The thermodynamic temperature scale was seen in Section 3.1 to be identical to the perfect gas scale. At normal temperature it is therefore possible to establish the thermodynamic scale by using a gas thermometer and applying a simple correction, such as the virial coefficient B_v in eqn 2.10, to the equation of state to allow for interactions between the molecules. It is possible to extend the scale down to about 1 K by measuring the saturated vapour pressure of ^{4}He and to 0.3 K using the rare isotope ^{3}He (Section 10.1).

In Section 10.4 however it was seen that ^{3}He could be cooled to 2×10^{-3} K (2 mK) using Pomeranchuk cooling and a similar temperature can be reached by the technique of adiabatic demagnetization discussed in the next section. The measurement of temperature in this region can be carried out most simply by measuring the susceptibility of either a nuclear paramagnet or a highly dilute electronic paramagnet. The susceptibility of a sphere of CMN, for example is found to follow Curie's law down to the mK region as would be expected from eqn 10.101 if only dipolar interactions are important.

A magnetic temperature T^* is commonly *defined* by the equation

$$\chi_{int}^{sphere} = \frac{C}{T^*} \tag{10.111}$$

where χ_{int}^{sphere} is the measured susceptibility of the sphere and by eqn 10.100, T^* is identical to T while Curie's law holds. At sufficiently low temperature it has already been remarked that Curie's law must break down but it is still possible to correct T^* to T if the specific heat of the substance has been measured (see Zemansky and Dittman 1981).

The use of a sphere of CMN has become standard in the very low-temperature region and the correction to T^* is only about 5 per cent at a temperature of 5 mK rising to 33 per cent near 3 mK. At still lower temperatures it is possible to use nuclear magnetism as an accurate thermometer since a nuclear paramagnet has been seen to follow Curie's law down to the region below 0.1 mK.

The weak nuclear paramagnetism can be separated from other magnetic effects in the solid by the method of nuclear magnetic resonance. The sample is placed in a uniform magnetic field $\mu_0 H^*$ and an alternating radio frequency field of frequency v, with its magnetic vector at right angles to $\mu_0 H^*$ is absorbed when the resonance condition

$$hv_0 = 2\mu\mu_0 H^* = \Delta$$

is satisfied.

This strength of the resonance signal depends upon the *difference* in the occupations of the two levels (for spin $\frac{1}{2}$) as was discussed in Section 7.8 so

$$\text{Signal} = \text{constant} \times N_s \left(\frac{1 - e^{-\beta \Delta}}{1 + e^{-\beta \Delta}} \right) \qquad (10.112)$$

$$= \frac{\text{constant}}{T} \qquad (\beta \Delta \ll 1)$$

provides a direct measurement of the thermodynamic temperature.

10.5.5 Adiabatic demagnetization

The attainment of temperatures in the mK region by the method of Pomeran-chuk cooling was discussed in Section 10.4. At such low temperatures it becomes difficult to establish thermal equilibrium between different parts of a system and the disadvantage of Pomeranchuk cooling was seen to be that solid ^{3}He forms in the *warmest* part of the system and therefore tends to isolate the sample from the liquid ^{3}He. Pomeranchuk cooling has therefore been largely restricted to the study of the properties of ^{3}He in the mK region since the sample then 'cools itself'.

The other important method of reaching the mK region was developed before Pomeranchuk cooling and is called, rather misleadingly, adiabatic demagnetization. The principle, as opposed to the practice, of the method is straightforward. The entropy of a system of particles with spin $\frac{1}{2}$ is simply given by the result for the two-level system, eqn 7.46

$$S_S = N_S k \left[\ln(1 + e^{-2x}) + \frac{2x}{1 + e^{2x}} \right] \qquad (10.113)$$

where

$$x = \frac{\mu \mu_0 H^{\text{loc}}}{kT}.$$

A reversible adiabatic (isentropic) decrease of the external magnetic field from $\mu_0 H_1^*$ to $\mu_0 H_2^*$ must take place at constant x and leads to a decrease in the spin temperature from T_1 to T_2 given by

$$\frac{H_1^{\text{loc}}}{T_1} = \frac{H_2^{\text{loc}}}{T_2} \qquad (10.114)$$

where H^{loc} is related to H^* by eqn 10.97. This equation may also be derived from thermodynamics, Exercise 10.9.

An external field as large as 5 T is now commonly available using a superconducting magnet, and reducing the current in the solenoid to zero would leave only the small field due to dipole–dipole interactions, say 0.01 T in a dilute electronic paramagnet, so the *spin* temperature (Section 7.8) could be reduced by a factor of 500 using this technique. The reduction factor would be even greater in a nuclear paramagnet where the dipolar field is only about 10^{-4} T.

The reason that the term 'adiabatic demagnetization' is rather misleading is that for an ideal paramagnet the magnetization of the system is constant in this process if $\mu_0 H_2^*$ is not reduced to zero since the magnetization is a function of H/T (eqn 10.91).

Equation 10.114 relates the initial and final *spin* temperatures of the system but is only correct for the final temperature of the whole system of lattice and spins if the entropy of the lattice is negligible. In general

$$\Delta S_S + \Delta S_L = 0 \tag{10.115}$$

for a reversible adiabatic decrease in the magnetic field, where S_L is the entropy of the lattice.

In an insulator at low temperature, from Section 10.3

$$\Delta S_1 = \int_{T_1}^{T_L} \frac{C_V}{T} dT = A \int_{T_1}^{T_L} T^2 dT = \frac{A}{3}(T_L^3 - T_1^3) \tag{10.116}$$

where T_1 is the initial temperature of spins and lattice and T_L is the final lattice temperature. The decrease in the lattice entropy must be balanced by increasing the spin temperature from the value T_2 (calculated using eqn 10.114) to T_L.

Although it is straightforward to write eqn 10.115 explicitly using eqns 10.113 and 10.116 the important condition for the difference $(T_L - T_2)/T_2$ to be small can be seen at once from the diagram for the spin entropy, Fig. 10.24. Unless the temperature can be lowered sufficiently for ΔS_L to be negligible

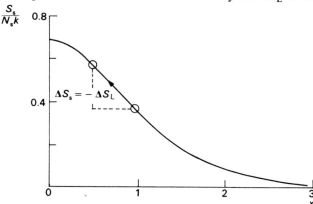

Fig. 10.24 The entropy of a simple paramagnet as a function of $x = \mu\mu_0 H^*/kT$. When the entropy of the lattice is negligible the entropy of the spin system is unchanged after adiabatic demagnetization and eqn 10.114 applies. When the entropy of the lattice cannot be ignored the spin entropy must increase to balance the decrease in lattice entropy at low temperature. (See Exercise 10.8)

relative to ΔS_S the spin system must initially be prepared on the steepest part of the spin entropy curve. A large change in spin entropy then leads to only a small change in temperature.

The condition is therefore

$$x = \frac{\mu\mu_0 H_1^*}{kT_1} \approx 1$$

where the difference between the external and local fields has been neglected.

A liquid helium cryostat using ^{3}He may be used to cool the paramagnet to 0.3 K in a field of 5 T but even in this most favourable case the value of x is only $\approx 10^{-2}$ for a nuclear paramagnet, so direct cooling from liquid helium temperatures to the mK range is not possible using nuclear demagnetization. In an electronic paramagnet such as CMN however it is simple to make $x \approx 1$ and in principle to obtain cooling to a few mK. A schematic diagram of the technique is shown in Fig. 10.25 and a numerical example will be found in Exercise 10.8. The experimental difficulties associated with the method are related to the small value of the heat capacity at such low temperature. A minute heat leak into the system or the smallest mechanical vibration will prevent the attainment of the temperature expected on the basis of eqn 10.109.

A further development of the method of adiabatic demagnetization has led to the lowest lattice temperatures yet attained. The idea is as simple as before but the experimental problems are far more formidable. In the technique of adiabatic double demagnetization (Fig. 10.25) an electronic paramagnet is first demagnetized and allowed to cool a nuclear paramagnet. The value of x in eqn 10.113 for the nuclear paramagnet might then become as large as 0.1 for a field of 5 T and a temperature of about 0.01 K. The thermal contact between the two paramagnets is now broken and the external field around the nuclear paramagnet reduced to zero. The nuclear spins are now in a local field of only about 10^{-4} T due to dipolar interactions and the final temperature may be as low as 1 μK. The temperature is measured using the susceptibility of the nuclear system.

A good account of the method of double demagnetization will be found in Zemansky and Dittman (1981). It should be noted that in practical calculations for electronic paramagnets such as CMN the simple spin-$\frac{1}{2}$ model is inadequate since the Cr^{3+} ion has non-zero angular momentum ($J = 5/2$) and the energy levels of the ion are affected by local *electric* fields in the crystal as well as by magnetic fields.

Exercises

10.1. Find expressions for the change in temperature of a slightly degenerate gas of fermions or bosons after a Joule–Kelvin expansion and explain the result. (Hint, use $PV = A + B_P P + \ldots$ and Exercise 2.2.)

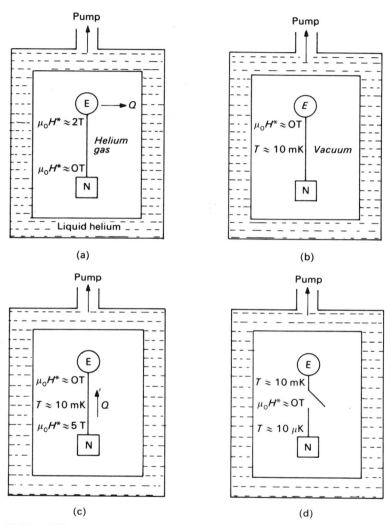

Fig. 10.25 (a) The electronic paramagnet (E) is magnetized at constant temperature (≈ 1 K). (b) The heat leak to the helium bath is broken and the external field reduced slowly to zero. The temperature of the system falls to a few mK. (c) The nuclear paramagnet (N) is now magnetized. The temperature of the combined magnetic system therefore increases slightly as may be seen from eqn 10.109. (d) Finally the nuclear paramagnet is isolated and demagnetised and the temperature falls to the μK region.

10.2. Find the pressure exerted by a gas of fermions at absolute zero at the same density as conduction electrons in silver.

10.3. A cavity contains a monatomic gas at a pressure of 1 atmosphere and black body radiation. Show that at room temperature the radiation makes a negligible contribution to the heat capacity of the system at constant volume.

10.4. Find an expression for the energy density of black body radiation as a function of wavelength and temperature $[u(\lambda, T)]$. Show $\lambda_m T$ is a constant and that $\lambda_m v_m \approx 3c/5$ where v_m is the peak of the $u(v, T)$ distribution.

10.5. Show that the lattice specific heat of an n-dimensional system is proportional to T^n at low temperature.

10.6. The dispersion relation for ferromagnetic excitations (spin waves or magnons) has the form $v \propto k^2$ for small k. Show that the three-dimensional magnon specific heat is proportional to $T^{3/2}$ at low temperature.

10.7. The spin wave dispersion relation in an antiferromagnet has the form $v \propto k$. What is the low-temperature dependence of the magnon specific heat?

10.8. The lattice heat capacity of a solid of N particles at low temperature may be written
$$C_L = \alpha N k T^3$$
where $\alpha \approx 10^{-5}\,\text{K}^{-3}$. Find the temperature at which the lattice and spin entropies are equal in a magnetic field such that x in eqn 10.113 is equal to one if N/N_S is equal to 60, where N_S is the number of particles with spin $\tfrac{1}{2}$. $(\partial S_S/\partial x) = -0.2\,N_S k$ near $x = 1$. Show that the final temperature of the whole system is twice that predicted by eqn 10.113 if the original lattice temperature is 10 K but that the lattice contribution is negligible for an original temperature of 1 K.

10.9. Show, using thermodynamics only, that $C_H = \mu_0 CH^{*2}/T^2$ for unit volume of an ideal paramagnet. Hence show that if other contributions to the heat capacity can be neglected: $H_2^*/T_2 = H_1^*/T_1$ for an isentropic process. (Hint: constant entropy is equivalent to constant magnetization in this case.)

10.10. The expression for the Fermi energy (eqn 10.12) depends upon the dimensions of the system. Show that the density of states for a two dimensional electron gas of surface area S is
$$D(\varepsilon)\,d\varepsilon = 2\pi g m S/h^2$$

i.e. independent of energy, and for N electrons,

$$\varepsilon_F = h^2 N / 4\pi S$$

10.11. Show that for a three-dimensional gas of free electrons: (a) $\partial(k_F r)/\partial V = 0$; (b) $\partial \ln \varepsilon_F / \partial \ln V = -2/3$, (c) $\partial \ln D(\varepsilon_F)/\partial \ln V = 2/3$.
The results are important for the study of the magnetic properties of e.g. Mn in Cu as a function of pressure.

10.12. Helium 3 is a fermi particle with spin$\frac{1}{2}$. Helium 4 is a bose particle as was discussed in Section 10.1. Dilute solutions of ^{3}He in liquid ^{4}He behave like a perfect Fermi gas except that the mass of the free atom (m) has to be replaced by an effective mass (m*). The Fermi temperature for a concentration of 1.3 per cent ^{3}He in liquid ^{4}He is found from specific heat measurements to be 0.14 K. The density of liquid ^{4}He is 145 kg m^{-3}. Show that m* = 2.3 m.

10.13. In a pioneering study of fluctuation phenomena and radiation Einstein showed that the relation derived in Exercise 10.4, $\lambda_m T = (hc/5k)$, meant that the r.m.s. flucatuation in the mean energy of the radiation in a volume λ_m^3 was of order $\bar{E}$. Show $\lambda_m T \simeq 0.3 \, hc/k$ from eqn 7.19.

10.14. The amplitude of lattice vibrations (u) is of importance to the understanding of X-ray scattering and the Mössbauer effect in solids. The zero point motion of a quantized simple harmonic oscillator, eqn 7.51, was neglected in Section 10.3, since it does not contribute to the heat capacity, but must be included in the analysis of lattice vibrations. (a) Show that for the three dimensional Einstein model of a solid at temperature T the displacement of an atom (u_T) is given by,

$$\tfrac{1}{2}\omega_E^2 m \overline{u_T^2} = \frac{\bar{\varepsilon}}{2}$$

$$\overline{u_0^2} = 3\hbar/2\omega_e m = 3\hbar^2/2mk\theta_E$$

and in the high temperature limit,

$$\overline{u_T^2} = 3kT/m\omega_E^2 = 3\hbar^2 T/mk\theta_E^2$$

where $\hbar\omega_E = h\nu_E = k\theta_E$.

(b) In the Debye model

$$\overline{u_T^2} = \frac{1}{3mN} \int_0^{\omega_D} \frac{\overline{\varepsilon(\omega)}}{\omega^2} D(\omega)\, d\omega.$$

Show that

$$\overline{u_T^2} = 3\hbar^2 T \left[\Phi(x) + x/4 \right]/mk\theta_D^2$$

where

$$\Phi(x) = \frac{1}{x} \int_0^x \frac{y}{e^y - 1} \, dy$$

and

$$x = \theta_D / T.$$

(c) Show that at 0 K

$$\overline{u_0^2} = 3\hbar^2 / 4mk\theta_D$$

and in the high temperature limit

$$\overline{u_T^2} = 3\hbar^2 T / mk\theta_D^2.$$

(d) In a lattice without zero point energy, if this were permitted by quantum mechanics, the atoms would be at rest at 0 K. Show that the actual amplitude of vibration of the atoms at 0 K is equivalent to that of a classical system at a temperature $\theta_D/4$. (e) Show that at the melting point of aluminium (933 K) the root mean square amplitude of vibration is approximately three times that at 0 K. Take $\theta_D = 438$ K. (See e.g. Willis and Pryor (1975) for further details).

10.15. A ferromagnetic material loses its long range order at a temperature called the ferromagnetic Curie point, T_c. Above T_c experimental data for the paramagnetic susceptibility is often plotted as $\ln \chi$ v. $\ln t$, where $t = (T - T_c)/T_c$, and a function $\gamma(t)$ defined by

$$\gamma(t) = -(\partial \ln \chi / \partial \ln t).$$

As discussed in Section 10.4 $\gamma(0)$ is a critical point exponent with a value of e.g. 1.35 for nickel. It is found that $\gamma(t)$ differs appreciably from $\gamma(0)$ unless $t \gtrsim 10^{-2}$. A new exponent, defined by the equation

$$\chi T \approx \left(\frac{T - T_c}{T} \right)^{-\hat{\gamma}(t)} \tag{E1}$$

has been shown to be nearly independent of temperature for some systems and may be taken to be a constant in the following calculations ($\hat{\gamma}$). (a) Show that eqn E1 may be written

$$\chi T_c \approx (1 + t)^{(\hat{\gamma} - 1)} t^{-\hat{\gamma}}$$

(b) Form $\partial \ln \chi / \partial \ln t$ and show that

$$\gamma(t) = (\hat{\gamma} + t)/(1 + t)$$

$$\gamma(0) = \hat{\gamma}; \quad \gamma(\infty) = 1.$$

Take $\hat{\gamma} = 1.35$ and plot $\gamma(t)$ against log t. (c) Show that for $T \gg T_c$ eqn E1 becomes

$$\chi^{-1} \approx T - T_P \tag{E2}$$

$$T_P = \hat{\gamma} T_c.$$

Sketch the form of $\chi^{-1} v \, T$ from eqns E1 and E2. T_P is the paramagnetic Curie point. See Fähnle and Souletie (1986) and Arrott (1984) and references there-in for further details.

11

Conclusion

The student who has reached this point in the book and has studied the exercises should now have some idea of the methods and range of application of thermal physics. A number of difficulties were, however, evaded in this book in the presentation of the foundations of thermodynamics and statistical mechanics partly in the interest of brevity and partly because these are best left until after the student has some familiarity with the subject. A number of books for further reading are listed after the appendixes but it may also be useful to discuss briefly some important topics which were not covered, either because of lack of space or because they were too difficult for an introductory text.

The extension of thermodynamics to systems with a variable number of particles (as in chemical processes) is not difficult and is discussed in Zemansky and Dittman (1981) and in many text books on chemical thermodynamics. The grand canonical ensemble was seen in Chapter 7 to be the appropriate ensemble for a system in thermal equilibrium but with a variable number of particles. In a large system the mean number of particles $\bar{N}$ is well defined and the grand canonical ensemble may then also be used to calculate the properties of a system with a constant number of particles ($N = \bar{N}$). There are often computational advantages in using the grand canonical ensemble rather than the canonical ensemble as may be seen in Kittel and Kroemer (1980), Huang (1963) or Reichl (1980).

The importance of fluctuations about equilibrium values in the canonical ensemble has been discussed several times in this book and particularly in connection with critical phenomena. A connection between the fluctuations of a system about equilibrium and its transport properties was first established by Einstein while studying the theory of Brownian motion. A particularly good discussion of this topic is to be found in Chapter 15 of Reif (1965). The general relation between the equilibrium and transport properties is discussed in Chapter 22 of Wannier (1966).

Classical thermodynamics has been seen to be restricted to systems in thermal equilibrium but a theory of non-equilibrium systems has also now been constructed. The theory of the transport properties of gases when the deviation from equilibrium was small was considered in terms of the kinetic theory in Chapter 9. The transport coefficients in this region are independent of the driving force, as in Ohm's law for example.

A much more difficult subject is the thermodynamics of systems far from equilibrium where the linear transport theory does not apply. Glansdorf and Prigorgine (1971) describe the progress in this field which has applications to many branches of science and possibly to the problem of evolution. Reichl (1980) also contains a useful account of this area.

One of the main interests of current research in statistical mechanics is in the behaviour of strongly interacting systems such as dense gases, liquids, and magnetic systems. The partition function of such systems can usually only be evaluated by approximate methods which require advanced mathematical techniques. A few models of strongly interacting systems can be solved exactly however as discussed in Huang (1963), Wannier (1966), and Reichl (1980).

Appendix I

Functions of two or more variables

When three variables are related by an equation of the form

$$z = f(x, y)$$

the total differential of z is defined to be

$$dz = \left(\frac{\partial z}{\partial x}\right)_y dx + \left(\frac{\partial z}{\partial y}\right)_x dy \qquad (A1.1)$$

where the suffix y in the first term on the right-hand side of the equation means that the differentiation with respect to x is to be performed with y treated as a constant. A term such as $(\partial z/\partial x)_y$ is called a partial differential. The equation could equally well be written in terms of x or y and for x would read

$$dx = \left(\frac{\partial x}{\partial y}\right)_z dy + \left(\frac{\partial x}{\partial z}\right)_y dz.$$

A number of relations exist between the partial differentials and the use of these relations forms an essential part of thermodynamics. Consider first the partial differential $(\partial x/\partial z)_y$. At constant y the second term in eqn A1.1 is zero so

$$\left(\frac{\partial x}{\partial z}\right)_y = \frac{1}{(\partial z/\partial x)_y}. \qquad (A1.2)$$

The differential of y with respect to x at constant z leads to

$$0 = \left(\frac{\partial z}{\partial x}\right)_y + \left(\frac{\partial z}{\partial y}\right)_x \left(\frac{\partial y}{\partial x}\right)_z$$

which may be written

$$\left(\frac{\partial z}{\partial y}\right)_x \left(\frac{\partial x}{\partial z}\right)_y \left(\frac{\partial y}{\partial x}\right)_z = -1. \qquad (A1.3)$$

The differential of z with respect to say x when some other variable ϕ is held constant leads to the equation

$$\left(\frac{\partial z}{\partial x}\right)_\phi = \left(\frac{\partial z}{\partial x}\right)_y + \left(\frac{\partial z}{\partial y}\right)_x \left(\frac{\partial y}{\partial x}\right)_\phi. \qquad (A1.4)$$

Finally, the order of differentiation is not important

$$\frac{\partial^2 z}{\partial x \partial y} = \frac{\partial^2 z}{\partial y \partial x}.$$ (A1.5)

As an example consider

$$z = x^2 y + xy^3$$

$$\left(\frac{\partial z}{\partial y}\right)_x = x^2 + 3y^2 x$$

$$\frac{\partial^2 z}{\partial x \partial y} = 2x + 3y^2.$$

Alternatively

$$\left(\frac{\partial z}{\partial x}\right)_y = 2xy + y^3$$

$$\frac{\partial^2 z}{\partial y \partial x} = 2x + 3y^2 = \frac{\partial^2 z}{\partial x \partial y}.$$

An expression of the form

$$A \, dx + B \, dy$$ (A1.6)

where A and B are functions of x and y is called a perfect (or exact) differential if some function z exists such that

$$dz = A \, dx + B \, dy.$$ (A1.7)

The test for a perfect differential is that

$$\left(\frac{\partial A}{\partial y}\right)_x = \left(\frac{\partial B}{\partial x}\right)_y$$ (A1.8)

since in this case A must be equal to $(\partial z/\partial x)_y$ and the equality in eqn A1.8 follows from eqn A1.5.

When the condition given by eqn A1.8 is not satisfied, eqn A1.6 is termed an imperfect differential. It may be shown that an integrating factor (λ) always exists for a function of two variables such that

$$\lambda A \, dx + \lambda B \, dy$$

is a perfect differential (where λ is some function of x and y) but for a function of *more* than two variables an integrating factor does not exist in general. The second law of thermodynamics expressed (in part) by the form 'an integrating factor always exists for the imperfect differential form of the first law of thermodynamics' is therefore a law of physics, not of mathematics.

As an example consider the expression

$$\text{d}L = xy^2\,\text{d}x + (x^2 y + xy)\text{d}y$$

where đ means that đL is an imperfect differential since

$$\left(\frac{\partial A}{\partial y}\right)_x = 2xy; \left(\frac{\partial B}{\partial x}\right)_y = (2x+1)y.$$

However

$$\text{d}z = \frac{\text{d}L}{xy} = y\,\text{d}x + (x+1)\text{d}y$$

is a perfect differential. In this case the integrating factor λ and function z are given by

$$\lambda = (xy)^{-1}; z = (x+1)y$$

The integrating factor is not however *unique* since the combination

$$\lambda^* = 2(x+1)/x; z^* = z^2$$

is also satisfactory. Notice that a suitable choice of z^* has led in this case to λ^* being a function of only *one* variable.

The fact that two functions (z, λ) are available explains why an integrating factor always exists for a function of two variables but for functions of three or more variables an integrating factor exists only for a restricted class of functions.

In the case of the differential form of the second law of thermodynamics discussed in Chapter 3

$$\text{d}S = \lambda\,\text{d}Q_R$$

The choice of the entropy (S) as a *extensive* quantity makes λ a function of only one variable, the temperature.

A fuller account of the relation between the mathematical and physical statements of Carathéodory than that given in Chapter 3 will be found in the book by Buchdahl (1966) quoted in the reading list.

Appendix II

Useful mathematics

Many of the integrals which occur in statistical physics may be reduced to standard forms. The gamma function is defined to be

$$\Gamma(n) = \int_0^\infty e^{-x} x^{n-1} \, dx \qquad (n > 0).$$ (A2.1)

Notice that the power of x is $(n-1)$ for $\Gamma(n)$. The following relationship exists between the gamma functions

$$\Gamma(n) = (n-1) \, \Gamma(n-1)$$ (A2.2)

and in particular

$$\Gamma(1) = 1$$ (A2.3)

$$\Gamma(\tfrac{1}{2}) = \sqrt{\pi}.$$ (A2.4)

An integral of the form

$$I(n) = \int_0^\infty e^{-\alpha x^2} x^n \, dx \qquad (n \geqslant 0)$$ (A2.5)

where α is a constant may be reduced to the form

$$I(n) = \tfrac{1}{2}\Gamma\left(\frac{n+1}{2}\right) \alpha^{-\frac{n+1}{2}}.$$ (A2.6)

An integral of the form

$$\int_{-\infty}^\infty e^{-\alpha x^2} x^n \, dx = 2I(n) \quad (n \text{ even})$$ (A2.7)

$$= 0 \quad (n \text{ odd}).$$ (A2.8)

The Riemann zeta function

$$\zeta(z) = \sum_{r=1}^\infty \frac{1}{r^z} \qquad (z > 1)$$

which has the values

$$\zeta(\tfrac{3}{2}) = 2.61 \quad \zeta(2) = \frac{\pi^2}{6} \quad \zeta(4) = \frac{\pi^4}{90}$$

is useful for the evaluation of the integral

$$\int_0^\infty \frac{x^{z-1}\,\mathrm{d}x}{e^x-1}=\zeta(z)\Gamma(z) \tag{A2.9}$$

Stirling's approximation

$$\ln n! = n\ln n - n + \tfrac{1}{2}\ln(2\pi n) \quad (n \gg 1)$$

where

$$n! = n(n-1)(n-2)\ldots 3\times 2\times 1$$

is accurate to better than 1 per cent for $n >$ ten. When n is very large the last term is negligible and the form

$$\ln n! = n\ln n - n \tag{A2.10}$$

is adequate (Chapter 6).

Appendix III

Lagrange undetermined multipliers

The total differential of a function of two variables (Appendix I) may be written

$$dz = \left(\frac{\partial z}{\partial x}\right)_y dx + \left(\frac{\partial z}{\partial y}\right)_x dy.$$

The condition for a maximum value of z is not only that dz is zero but that each term is zero.

$$\left(\frac{\partial z}{\partial x}\right)_y = \left(\frac{\partial z}{\partial y}\right)_x = 0 \tag{A3.1}$$

since x and y are independent variables.

However when a maximum value of z subject to a *constraint* of the form

$$f(x, y) = \text{constant} \tag{A3.2}$$

is required, x and y are no longer independent since

$$df = \left(\frac{\partial f}{\partial x}\right)_y dx + \left(\frac{\partial f}{\partial y}\right)_x dy = 0$$

but

$$\frac{(\partial z/\partial x)_y}{(\partial f/\partial x)_y} = \frac{(\partial z/\partial y)_x}{(\partial f/\partial y)_x}.$$

The equations may now be written in terms of the (unknown) constant (λ)

$$\left(\frac{\partial z}{\partial x}\right)_y - \lambda\left(\frac{\partial f}{\partial x}\right)_y = 0$$

$$\left(\frac{\partial z}{\partial y}\right)_x - \lambda\left(\frac{\partial f}{\partial y}\right)_x = 0$$

These equations would result if the function

$$\phi(x, y) = z - \lambda f(x, y)$$

were maximized *without* a constraint, as in eqn A3.1, so the effect of the constant λ (the Lagrange undetermined multiplier) is to decouple the equations. The complete solution of the maximum, subject to the restraint, has of course only been postponed since λ is unknown but λ may be carried to the end of the problem and then adjusted to give the correct value of the constant in eqn A3.2.

The extension of the method when z is a function of n independent variables subject to m constraints $(m < n)$ is straightforward since one undetermined multiplier is simply introduced for each constraint and the terms become of the form

$$\left(\frac{\partial z}{\partial x}\right)_{x'} - \lambda_1\left(\frac{\partial f_1}{\partial x}\right)_{x'} - \lambda_2\left(\frac{\partial f_2}{\partial x}\right)_{x'} \cdots - \lambda_m\left(\frac{\partial f_m}{\partial x}\right)_{x'} = 0$$

where the suffix x' means that all variables apart from x are to be held constant.

Appendix IV

Density of single-particle translational states

The allowed solutions of any wave equation depend upon the boundary conditions imposed on the wave. The most familiar example is perhaps that of a stretched wire held at both ends. The only standing transverse waves which can be excited in the wire are those which have zero amplitude at both ends. This restriction means that the wavelengths of the standing waves must be given by an equation of the form

$$n\frac{\lambda}{2} = L \quad (n = 1, 2, 3, \ldots) \tag{A4.1}$$

where L is the distance between the ends of the wire.

The wave vector (k) is defined by

$$k = \frac{2\pi}{\lambda} \tag{A4.2}$$

so the allowed wave vectors are given by

$$k = \frac{\pi n}{L} \quad (n = 1, 2, 3, \ldots). \tag{A4.3}$$

The allowed wave vectors may be represented by a set of points spaced at intervals of π/L. The space so defined (in this one-dimensional case simply a line) is called k-space and has the dimensions of inverse length.

The concept of discrete allowed solutions to a wave equation is not therefore in any way restricted to quantum mechanics but became of greatly increased importance after the introduction of the Schrödinger wave equation. The case of a particle in a box will now be considered in detail and then the general results for different kinds of system will be discussed.

In one dimension the Schrödinger equation for a free particle of mass m may be written

$$\frac{\mathrm{d}^2\psi}{\mathrm{d}x^2} + \frac{8\pi^2 m\varepsilon}{h^2}\psi = 0 \tag{A4.4}$$

The solution to this equation is of the form

$$\psi = A\mathrm{e}^{ik_x x} \tag{A4.5}$$

$$\varepsilon = \frac{h^2 k_x^2}{8\pi^2 m} = \frac{p_x^2}{2m} \tag{A4.6}$$

$$p_x = \frac{h}{\lambda} = \frac{h}{2\pi} k_x \tag{A4.7}$$

where A is a normalising constant, ε the energy of the particle, and p_x the momentum in the x-direction. The last equation is simply the de Broglie relationship between the momentum and wavelength of a particle.

When the particle is free to move only within the confines of a box (or in the present one-dimensional case along a line of length L) the wave vector must again be restricted to those values which satisfy the boundary conditions. The boundary conditions are less obvious than for the case of the stretched string but when the box is large the final results for the macroscopic properties of the system cannot depend upon the exact size of the box or the choice of boundary condition. One possible boundary condition is that the wave function vanish at the surface of the box. This is equivalent to the standing wave solution of eqn A4.1.

An alternative boundary condition, which will be used throughout the book, is found by joining the two ends of the length L and considering *running* waves around the loop. This may be expressed

$$\psi(x) = \psi(x + L) \tag{A4.8}$$

and is known as a periodic boundary condition. The allowed wavelengths are now

$$n\lambda = L \qquad (n = 0, \pm 1, \pm 2, \pm \dots) \tag{A4.9}$$

$$k_x = \frac{2\pi n}{L} \qquad (n = 0, \pm 1, \pm 2, \dots). \tag{A4.10}$$

The allowed energies and momenta of the particle may now be found by substituting for k_x in eqns A4.6 and A4.7. The energy levels for periodic boundary conditions in one dimension are all (except for the ground state in which n equals zero) two-fold degenerate since the states with $\pm n$ have the same energy.

When L is large the allowed values of k (the microstates, or states of the system, since they represent the solutions of the Schrödinger equation) are sufficiently close together to be represented by a continuous function called the density of states. The exact meaning of 'sufficiently close' will be discussed after the results for three dimensions have been obtained.

The number of allowed states per unit length of the line in k-space is, from eqn A4.10, equal to $L/2\pi$.

The density of states function in one dimension is defined as

$$D(k)\,dk = \text{number of states in the range } k \text{ to } k + dk$$

$$= \frac{L}{2\pi}\,dk.$$

When the particle is restricted to a two-dimensional box of side L the allowed states for periodic boundary conditions are given by

$$k_x = \frac{2\pi n_x}{L} \quad (n_x = 0, \pm 1, \pm 2, \pm \ldots) \tag{A4.12}$$

$$k_y = \frac{2\pi n_y}{L} \quad (n_y = 0, \pm 1, \pm 2, \pm \ldots) \tag{A4.13}$$

and the allowed energy levels by

$$\varepsilon = \frac{h^2(k_x^2 + k_y^2)}{8\pi^2 m} = \frac{h^2}{2mL^2}(n_x^2 + n_y^2). \tag{A4.14}$$

The energy levels above the ground state can be more degenerate than in one dimension because the same energy can be obtained, for example, with either n_x equal to ± 2 and n_y equal to ± 1 or vice versa.

In two (and three) dimensions the term 'density of states' is commonly used for two rather different quantities and this sometimes leads to confusion. The allowed states in two dimensions form a square array in k-space with side $2\pi/L$ (Fig. A4.1).

The number of states in a small element of area $d^2 k$ is

$$\left(\frac{L}{2\pi}\right) d^2 k \tag{A4.15}$$

where the symbol $d^2 k$ is used to stress that k is a vector quantity. The density of states function in this sense is a constant throughout all the k-space as may be seen from Fig. A4.1.

The alternative use of the term 'density of states' refers to the *magnitude* of k (a scalar quantity). $D(k)\,dk$ is the number of states with wave vectors of magnitude between k and $k + dk$. The wave vectors of constant magnitude k lie on a circle about the origin in k-space (Fig. A4.1). The area between vectors of magnitude k and $k + dk$ is simply $2\pi k\,dk$ and therefore

$$D(k)\,dk = \left(\frac{L}{2\pi}\right)^2 2\pi k\,dk = \frac{L^2}{2\pi}\,k\,dk. \tag{A4.16}$$

When scalar quantities (such as energy) are under consideration it is this density of states function which is of importance. The relationship between

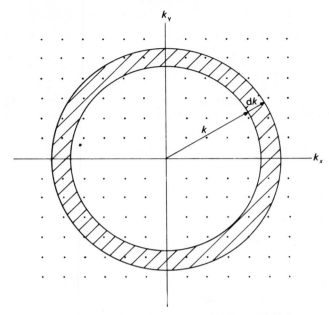

Fig. A4.1   The allowed single-particle states in a two-dimensional box of side L form a square array in k-space with spacing $2\pi/L$. The allowed states with wave vector of *magnitude* between k and $k+\mathrm{d}k$ lie in the shaded area.

eqn A4.15 and A4.16 may be clarified if the area element of $\mathrm{d}^2\boldsymbol{k}$ is written in polar coordinates. Then eqn A4.15 becomes

$$\left(\frac{L}{2\pi}\right)^2 k\,\mathrm{d}k\,\mathrm{d}\phi. \qquad (A4.17)$$

The scalar density of states therefore results from integrating eqn A4.17 over the whole angular range 0 to 2π at constant wave vector.

   The extension of these concepts to three dimensions is straightforward. The periodic boundary conditions may still be used although in fact the opposite faces of a cube can only be joined together in a four-dimensional space. The energy of the particle is given by the obvious generalization of eqn A4.14

$$\varepsilon_n = \frac{h^2(k_x^2 + k_y^2 + k_z^2)}{8\pi^2 m} = \frac{h^2 n^2}{2mL^2} \qquad (A4.18)$$

$$n^2 = n_x^2 + n_y^2 + n_z^2. \qquad (A4.19)$$

The energy ε_n depends only upon n^2, but the degeneracy of an energy level depends upon the number of ways n^2 can be formed from values of n_x, n_y and n_z in eqn A4.18.

The allowed states form a cubic array in the three-dimensional k-space with spacing $2\pi/L$. The density of states may be written either as

$$\left(\frac{L}{2\pi}\right)^3 d^3\boldsymbol{k} \tag{A4.20}$$

if the vector quantity is under consideration (where $d^3\boldsymbol{k}$ may be written in any convenient form such as

$$d^3\boldsymbol{k} = dk_x dk_y dk_z \tag{A4.21}$$

in cartesian coordinates or

$$d^3\boldsymbol{k} = k^2 \, dk \sin\theta \, d\theta \, d\phi \tag{A4.22}$$

in polar coordinates) or as the scalar quantity defined in eqn A4.11.

The volume enclosed between the spheres of radius k and $k + dk$ is $4\pi k^2 \, dk$ so the scalar density of states in three dimensions is

$$D(k) \, dk = \left(\frac{L}{2\pi}\right)^3 4\pi k^2 \, dk = \frac{V}{2\pi^2} k^2 \, dk \tag{A4.23}$$

which, as in the two-dimensional case, could have been obtained by integrating over the angular part of eqn. A4.20 at constant k.

Although eqn A4.23 was derived for the specific case of a cube it is in fact correct for any large volume V. The density of states function given by eqn A4.23 is also obtained when the standing-wave boundary conditions (eqn A4.3) are used. In this case the points are twice as close together in k-space but the allowed values are restricted to one octant of the sphere (since all the components of k must be positive).

The physical meaning of the density of states function may be made apparent by considering the states in a space of n dimensions of momentum and n of position where n is 1, 2 or 3. This space is called a phase space. The wave vector and the momentum of the particle are related by eqn A4.7 so eqn A4.20 may be written

$$\left(\frac{L}{h}\right)^3 d^3\boldsymbol{p} \tag{A4.24}$$

or generalizing the volume term

$$\frac{d^3\boldsymbol{p}\,d^3\boldsymbol{r}}{h^3}. \tag{A4.25}$$

The meaning of eqn A4.25 may be expressed in words as 'a single-particle quantum state occupies a volume h^n in a phase space of $2n$ dimensions'.

This fundamental result is obviously a feature of the Heisenberg uncertainty principle and illustrates the difference between classical physics (which places no restriction on the spacing of the states) and quantum physics.

The most general form of the density of states function for a particle is therefore given by eqn A4.25. When only the scalar density of states is required and gravitational effects are neglected this becomes (from eqn A4.23)

$$D(p)\mathrm{d}p = \frac{4\pi V p^2}{h^3}\,\mathrm{d}p \qquad (A4.26)$$

using the mathematical relationship

$$D(p)\mathrm{d}p = D(k)\left(\frac{\mathrm{d}k}{\mathrm{d}p}\right)\mathrm{d}p \qquad (A4.27)$$

where the meaning of $D(p)$ is

$D(p)\mathrm{d}p = $ Number of states with magnitude of momentum

between

p and $p + \mathrm{d}p$. \qquad (A4.28)

The density of states function can also be expressed as a function of the energy of the particle provided that the relationship between the energy and the momentum is known. In Newtonian mechanics the energy and momentum are related by the equation

$$\varepsilon = \tfrac{1}{2}mv^2 = \frac{p^2}{2m}. \qquad (A4.29)$$

The density of states as a function of energy may then be written (using eqn A4.26)

$$D(\varepsilon)\,\mathrm{d}\varepsilon = \frac{4\pi V}{h^3}\,2m\varepsilon\left(\frac{m}{p}\right)\mathrm{d}\varepsilon \qquad (A4.30)$$

$$= 2\pi V\left(\frac{2m}{h^2}\right)^{\frac{3}{2}}\varepsilon^{1/2}\,\mathrm{d}\varepsilon. \qquad (A4.31)$$

A particle in the extreme relativistic limit however has the energy momentum relation

$$\varepsilon = cp \qquad (A4.32)$$

where c is the velocity of light. It is still permissible to use eqn A4.23 or eqn A4.26 (although eqn A4.30 does not apply) and therefore

$$D(\varepsilon)\,\mathrm{d}\varepsilon = \frac{4\pi V}{(hc)^3}\varepsilon^2\,\mathrm{d}\varepsilon. \qquad (A4.33)$$

Notice that this equation cannot be obtained from eqn A4.30. The basic eqns are A4.23 and A4.26.

When the particle has spin (s) each energy level, in the absence of a magnetic field, has an increased degeneracy since ($2s + 1$) states have the same kinetic energy. The density of states function, such as equation A4.31, must then be multiplied by $(2s+1)$. An electron for example has $s=1/2$, $(2s+1)=2$. The form of eqn A4.31 is correct for one direction of spin or for a particle with zero spin. The degeneracy factor will be written g where g is equal to $(2s+1)$ for particles with spin s, to 2 for electromagnetic waves and to 3 for lattice waves, as will be seen below.

The results of this section for the Schrödinger wave equation of a particle can be carried over to other wave equations using the basic result of eqn A4.23. An electromagnetic wave in a cavity for example (Section 10.2) has the same energy density of states as an electron in the extreme relativistic limit which is apparent from treating the electromagnetic field in terms of a photon gas since photons obey eqn A4.32. The degeneracy factor is again 2 due to the two directions of polarization of the wave.

A more complicated situation arises in the theory of the lattice vibrations of a solid (Section 10.3). The total number of allowed modes of the system is now *limited* to $3N$ for N atoms in a three-dimensional crystal. The exact treatment of this problem will not be considered here but in the extremely useful Debye approximation the density of states may be found immediately from eqn A4.23. In the Debye approximation the N atoms of the crystal are considered to form a continuum for which the frequency (v) and wave vector are related by

$$c_s = v\lambda = \frac{2\pi v}{k} \tag{A4.34}$$

where c_s is the velocity of sound in the crystal. The density of lattice vibrational states as a function of frequency is, from eqn A4.23

$$D(v)\,dv = 3 \times \frac{4\pi V}{c_s^3} v^2\,dv \tag{A4.35}$$

but the number of states must be restricted to $3N$ so there is a maximum value of the frequency (v_D) given by

$$\int_0^{v_D} D(v)\,dv = 3N. \tag{A4.36}$$

Therefore

$$v_D = \left(\frac{3Nc_s^3}{4\pi V}\right)^{\frac{1}{3}}. \tag{A4.37}$$

The density of states in eqn A4.35 includes the three possible degrees of polarization of the lattice vibrations.

The density of states function was stated to be a good approximation to the actual discrete allowed energy states provided that the states were 'sufficiently close together'. In the particular case of a particle in a box the meaning of 'sufficiently close' may be illustrated in terms of eqn A4.18.

The mean kinetic energy of a single particle in a box at temperature T is $3kT/2$ (Section 7.5). This defines a value of n_0 from eqn A4.18

$$n_0^2 = \frac{2mL^2\varepsilon}{h^2} = \frac{3mL^2kT}{h^2}.$$
(A4.38)

The separation of adjacent energy levels is

$$\Delta\varepsilon = \frac{h^2}{2mL^2}(2n_0 - 1) \approx \frac{h^2 n_0}{mL^2}$$
(A4.39)

which may be written in temperature units by defining a temperature T_g by the equation

$$\Delta\varepsilon = kT_g.$$
(A4.40)

An electron in a box of side 10^{-2} m at 300 K then has

$$n_0 \approx 10^5 \quad T_g \approx 10^{-5} \text{ K} \ll T$$
(A4.41)

so the separation between the energy levels is many orders of magnitude less than the thermal energy.

The density of states function may always be used in the classical limit and also for the Fermi–Dirac gas but has to be modified for the ground state of a Bose–Einstein gas in the low-temperature limit (Section 10.1).

Appendix V

Approximate values for fundamental constants

Charge on proton (e)	1.6×10^{-19} C
Planck constant (h)	6.6×10^{-34} J s
Avogadro constant (N_A)	6.0×10^{23} mol^{-1}
Gas constant (R)	8.3 J K^{-1} mol^{-1}
Boltzmann constant (k)	1.4×10^{-23} J K^{-1}
	8.6×10^{-5} eV K^{-1}
Stefan–Boltzmann constant (σ)	5.7×10^{-8} W m^{-2} K^{-4}
Rest mass of electron	9.1×10^{-31} kg
Rest mass of proton	1.7×10^{-27} kg
Rydberg constant (R_∞)	1.1×10^7 m^{-1}
Bohr magneton	9.3×10^{-24} J T^{-1}
Nuclear magneton	5.0×10^{-27} J T^{-1}
Permeability of free space (μ_0)	$4\pi \times 10^{-7}$ Hm^{-1}

A pressure of 1 atmosphere $\equiv 760$ mm mercury $= 1.0 \times 10^5$ N m^{-2} $= 1.0 \times 10^5$ Pa

(1 mm mercury $= 1$ torr $= 133$ Pa)

An energy of 1 eV $\equiv 1.6 \times 10^{-19}$ J

At room temperature a thermal energy $kT \approx 300$ K ≈ 0.026 eV

Normal temperature and pressure (NTP) $T = 273$ K, $P = 1$ atmosphere

One mole of a perfect gas occupies 22.41 (0.0224 m^3) at NTP

Appendix VI

Properties of the elements (required for exercises)

Element	Symbol	Atomic weight (g mol^{-1})	Atoms/molecule
Hydrogen	H	1	2
Deuterium	D	2	2
Helium*	He	3	1
Helium*	He	4	1
Nitrogen	N	14	2
Oxygen	O	16	2
Gallium	Ga	31	1
Argon	A	40	1

* Normal helium gas may be taken as ^{4}He.

N.B. In calculations using S.I. units the above values for atomic weight must be multiplied by 1×10^{-3} to convert them to kg mol^{-1}.

Appendix VII

Properties of gases at NTP*

	d† (nm)	$\bar{v}$ (ms^{-1})	λ† (nm)	η (μNsm^{-2})	κ (mWm^{-1} K^{-1})
Hydrogen	0.30	1,700	110	8.4	170
Helium	0.26	1,200	170	19	140
Nitrogen	0.37	450	59	17	24
Oxygen	0.36	430	63	20	25
Argon	0.34	380	63	21	16

* to 2 Significant figures.
† Values for the molecular diameter and mean free path depend slightly upon the method of calculation.

Further reading

General texts

A number of books cover much the same material as this book in greater detail and at a more advanced level. The full titles are listed in alphabetical order by author in the reference section.

Kittel and Kroemer (1980), Mandl (1971), Reif (1965) and Wannier (1966) all develop thermodynamics and statistical mechanics in parallel rather than (as in this book) as separate approaches to the study of systems in thermal equilibrium. Mandl (1971) is the least advanced of the four and is restricted to system in thermal equilibrium. Kittel and Kroemer (1980) begins by discussing a model systems in the same spirit as Section 6.3. They then go on to cover many interesting applications of statistical mechanics using the grand canonical ensemble but the foundations of statistical mechanics are not considered in any detail. Reif (1965) is an extremely detailed text, to the earlier sections of which this book may serve as a guide. The final chapters on transport theory and fluctuations are outstandingly good. Wannier (1966) also covers both equilibrium and transport theory and contains some unusual examples as well as a careful discussion of the relaxation-time approach to the Boltzmann equation. A very comprehensive account of statistical mechanics, at an advanced level, is given by Reichl (1980).

Thermodynamics

Zemansky and Dittman (1981) is not more advanced than this book but is useful for its very clear presentation and many tables and diagrams of experimental results. A good advanced text with an approach quite different to that adopted here is Callen (1960). The foundations of thermodynamics are discussed with great clarity in Buchdahl (1960). A full mathematical treatment of the Carathéodory approach to the second law of thermodynamics is given and comparison made with the earlier versions of the law due to Clausius and Kelvin. (See also Landsberg 1983).

Statistical mechanics

The classic text on the foundations of statistical mechanics is Tolman (1938–reprinted many times). A good account is also given in Huang (1963).

In addition to the books listed under 'general texts' one other outstanding advanced book on statistical mechanics is that by Landau and Lifshitz (1959). The fundamental properties of entropy, and its relationship to "information" are discussed in Denbigh and Denbigh (1985).

The thermodynamics of a system with a negative temperature was first discussed by Ramsey (1956). The concept of negative temperature is also used on books on masers and lasers, as for example Unger (1970).

Transport theory

Reif (1965), Wannier (1966), and Reichl (1980) contain good discussions of transport theory. A detailed review of the solution of the Boltzmann transport equation is given in Hirschfelder *et al.* (1954) but some of the more advanced sections of this book on dense gases and liquids are now out of date. The obvious applications of kinetic theory to vacuum physics was remarked on in Chapters 8 and 9. A useful introduction to ultra-high-vacuum techniques and interaction of the atoms of a gas with a surface is given in Weston (1985). Ziman (1960) contains a good discussion of the use of the Boltzmann transport equation in the theory of solids.

Chapter 10

10.1 Kittel and Kroemer (1980) and Huang (1963) give good discussions of Fermi–Dirac and Bose–Einstein statistics. Kittel (1986) is an excellent introduction to the properties of electrons in solids. The properties of liquid helium are discussed in Tilley and Tilley (1986).

10.2 The black body background radiation is discussed in a book by Sciama (1971) and in an article by Webster (1974). Experimental details are given in Witebsky *et al.* (1986).

10.3 Kittel (1986) and Gopal (1966) contain theory and comparison with experiment.

10.4 The early theory of critical phenomena is covered in a book by Stanley (1971). The modern theory is given in Reichl (1980). Careri (1984) contains an enjoyable descriptive account of phase transitions. Pomeranchuk cooling is discussed by Huiskamp and Lounasmaa (1973) and the superfluid properties of ^{3}He in the temperature range below 3 mK by Lounasmaa (1974).

10.5 A good account of the thermodynamics of magnetic systems is given in Kittel and Kroemer (1980) and, at a more advanced level, in Robinson (1973).

Additional reading

Buchdahl, H. A. (1966). *The concepts of classical thermodynamics*. Cambridge University Press, Cambridge.

Callen, H. B. (1960). *Thermodynamics*. Wiley, New York.

Careri, G. (1984). *Order and disorder in matter*. Benjamin/Cummings, Menlo Park, CA.

Denbigh, K. G. and Denbigh, J. S. (1985). *Entropy in relation to incomplete knowledge*. Cambridge University Press, Cambridge.

Glansdorf, P. and Prigorgine, I. (1971). *Thermodynamic theory of structure, stability and fluctuations*. Wiley Interscience, Chichester.

Gopal, E. S. R. (1966). *Specific heats at low temperature*. Heywood, London; Plenum Press, New York.

Hirschfelder, J. O., Curtiss, C. F., and Bird, R. B. (1954). *Molecular theory of gases and liquids*. Wiley, New York.

Huang, K. (1963). *Statistical mechanics*. Wiley, New York.

Huiskamp, W. J. and Lounasmaa, O. (1973). Ultralow temperatures—how and why. *Reports on Progress in Physics* **36**, 423–96.

Kittel, C. and Kroemer, H. (1980). *Thermal physics*. W. H. Freeman, San Franscisco.

Kittel, C. (1986). *Introduction to solid state physics* (6th edn.). Wiley, New York.

Klein, M. J. (1974). Carnot's contribution to thermodynamics. *Physics Today* **7**, 23–8 (August).

Kuhn, T. S. (1978). *Blackbody theory and the quantum discontinuity*. Oxford University Press, Oxford.

Landau, L. D. and Lifshitz, E. M. (1959). *Statistical physics*. Pergamon, London.

Landsberg, P. T. (1983). The Born Centenary: Remarks about classical thermodynamics. *American Journal of Physics* **51**, 842–5.

Lounasmaa, O. V. (1974). The superfluid phases of liquid ^{3}He. *Contemporary Physics* **15**, 353–74.

Mandl, F. (1971). *Statistical Physics*. Wiley, New York.

Ramsey, N. F. (1956). Thermodynamics and statistical mechanics at negative absolute temperatures. *Physical Review* **103**, 20–8.

Reichl, L. E. (1980). *A modern course in statistical physics*. Arnold, London.

Reif, F. (1965). *Statistical and thermal physics*. Wiley, New York.

Robinson, F. N. H. (1973). *Macroscopic electromagnetism*. Pergamon, Oxford.

Sciama, D. W. (1977). *Modern cosmology*. Cambridge University Press, Cambridge.

Stanley, H. E. (1971). *Introduction to phase transitions and critical phenomena*. Clarendon Press, Oxford.

Tilley, D. R. and Tilley, J. (1986). Superfluidity and superconductivity. Adam Hilger, Bristol.

Tolman, R. C. (1938). *The principles of statistical mechanics*. Oxford University Press, Oxford.

Unger, H. G. (1970). *Introduction to quantum electronics*. Pergamon, London.

Wannier, G. H. (1966). *Statistical physics*. Wiley, New York.

Webster, A. (1974). The cosmic background radiation. *Scientific American* **231**, 26–33 (August).

Weston, G. F. (1985). *Ultrahigh vacuum practice*. Butterworths.

Zemansky, M. W. and Dittman, R. H. (1981). *Heat and Thermodynamics* 6th edn. McGraw-Hill, New York.

Zman, J. M. (1960). *Electrons and Phonons*. Oxford University Press, Oxford.

References

Arrott, A. S. (1984). Analysis of high temperature susceptibility in iron and nickel. *Journal of Magnetism and Magnetic Materials* **45**, 59–66.

Berman, R., Foster, E. L., and Ziman, J. M. (1955). Thermal conduction in artificial sapphire crystals at low temperature. *Proceedings of the Royal Society* **231**, 130–44.

Brewer, D. F. (1970). Some thermal, magnetic and flow properties of adsorbed He and ^{3}He–^{4}He mixtures. *Journal of Low Temperature Physics* **3**, 205–24.

Caplin, A. D., Grüner, G., and Dunlop, J. B. (1973). $Al_{10}V$: An Einstein solid. *Physical Review Letters* **30**, 1138–40.

Carnot, S. (1824). Réflexions sur la puissance motrice de feu, Paris. [An English translation is available: (1960) (ed. E. Mendoza). Dover Publications, New York.]

Fähnle, M. and Souletie, J. (1986). The Generalized Curie–Weiss Law. *Physica status solidi* **138**, 181–8.

Fisher, M. E. (1964). Correlation functions and the critical region of simple fluids. *Journal of Mathematical Physics* **5**, 944–62.

Gibbs, J. W. (1902). *Elementary principles in statistical mechanics*. Yale University Press. [Reprinted (1960) Dover Publications, New York.]

Guillermet, A. F. (1986). The pressure dependence of the expansivity and of the Anderson–Grüneisen parameter in the Murnaghan approximation. *Journal of Physics and Chemistry of Solids* **47**, 605–7.

Luscher, P. E. and Collins, D. M. (1979). Design consideration for molecular beam epitaxy systems. *Progress in Crystal Growth Characterization* **2**, 15–32.

Miller, R. C. and Kusch, P. (1955). Velocity distributions in potassium and thallium atomic beams. *Physical Review* **99**, 1314–21.

Powles, J. G. (1983). The Boyle line. *Journal of Physics C: Solid State* **16**, 503–14.

Powles, J. G. (1968). *Particles*. Addison-Wesley, London.

Rushbrooke, G. S. (1963). On the thermodynamics of the critical region for the Ising problem. *Journal of Chemical Physics* **39**, 842–3.

Silvera, I. F. and Walraven, J. (1982). The stabilization of atomic hydrogen. *Scientific American* **246**, 56–71 (January).

Willis, B. T. M. and Pryor, A. W. (1975). *Thermal vibrations in crystallography*. Cambridge University Press, Cambridge.

Witebsky, C., Smoot, G., De Amici, G., and Friedman, S. D. (1986). New measurements of the cosmic background radiation at 3.3 millimeter wavelength. *Astrophysical Journal* **310**, 145–59.

Solutions to exercises

2.1. The Celsius temperature on the resistance scale is defined

$$\theta_R \equiv \frac{R_\theta - R_0}{R_{100} - R_0} \times 100\,°C$$

(since ice and steam points have the same value on all scales). At 70°C on the gas scale

$$R_\theta = R_0(1 + 70\alpha + 70^2\beta).$$

Therefore

$$\theta_R = (70\alpha + 70^2\beta)/(\alpha + 100\beta) = 72°C.$$

2.2.(a) At low density the equations must reduce to the perfect gas law. Now

$$(PV)_{\to A} V_{\to \infty}; (PV)_{\to A} P_{\to 0}$$

$$A = RT$$

(b), (c)

$$P = \frac{A}{V}\left(1 + \frac{B_V}{V} + \frac{C_V}{V^2} + \dots\right)$$

$$= \frac{A}{V} + \frac{AB_P}{V^2}\left(1 + \frac{B_V}{V} + \frac{C_V}{V^2} + \dots\right) + \frac{A^2 C_P}{V^3}\left(1 + \frac{B_V}{V} + \frac{C_V}{V^2} + \dots\right).$$

Equating coefficients of $1/V^2$

$$B_P = B_V$$

Equating coefficients of $1/V^3$

$$C_V = B_P^2 + AC_P$$

(d), (e)

$$P = \frac{RT}{V - b} - \frac{a}{V^2}$$

$$= \frac{RT}{V}\left(1 + \frac{b}{V} + \frac{b^2}{V^2} + \dots\right) - \frac{a}{V^2} \qquad (b \ll V).$$

Equating coefficients of $1/V^2$

$$B_V = b - a/RT$$

Equating coefficients of $1/V^3$

$$C_V = b^2$$

2.3.$(\partial P/\partial V)_T = 0$ at critical point
therefore

$$P = a(V - 2b)/V^3$$

On differentiating again with respect to V and equading to zero

$$V_c = 3b$$

substituting for V in the first equation then leads to

$$P_c = a/27b^2$$

and substituting the van der Waals equation with these values gives the other
equations.

2.4(a) $$W = -\int_{V_1}^{V_2} P dV = -P(V_2 - V_1) \text{ for } P \text{ constant}$$

(the sign convention is such that W is positive if work is done *on* the gas, that
is to say when V_2 is less than V_1).

(i) $$W = -R(T_2 - T_1)$$
(ii) $$W = -R(T_2 - T_1)$$

(iii) $$W = -R(T_2 - T_1) - a\left(\frac{1}{V_1} - \frac{1}{V_2}\right)$$

(iv) $$W = -R(T_2 - T_1) - a\left(\frac{1}{V_1} - \frac{1}{V_2}\right) + ab\left(\frac{1}{V_1^2} - \frac{1}{V_2^2}\right)$$

In eqn iv the term in 'a' represents the work done to overcome the attractive
force between the molecules. The term in 'b' represents the repulsive forces.
Notice that for expansion at constant pressure the repulsive term is not
important.

2.4.(b)

(i) $$W = -\int_{V_1}^{V_2} P dV = -RT \int_{V_1}^{V_2} \frac{dV}{V} = -RT \ln\left(\frac{V_2}{V_1}\right)$$

(ii) $$W = -RT \ln\left(\frac{V_2 - b}{V_1 - b}\right)$$

$$= -RT \ln\left(\frac{V_2}{V_1}\right) - RTb\left(\frac{1}{V_1} - \frac{1}{V_2}\right)$$

(iii) $$W = -RT \ln\left(\frac{V_2}{V_1}\right) + a\left(\frac{1}{V_1} - \frac{1}{V_2}\right)$$

(iv) $\qquad W = -RT \ln\left(\dfrac{V_2 - b}{V_1 - b}\right) + a\left(\dfrac{1}{V_1} - \dfrac{1}{V_2}\right)$

$$= -RT \ln\left(\dfrac{V_2}{V_1}\right) - Rb(T - T_B)\left(\dfrac{1}{V_1} - \dfrac{1}{V_2}\right)$$

Notice that the second equation of state led to the same work-term as a perfect gas in part (a) but not in part (b). Similarly the van der Waals gas would behave as a perfect gas at the Boyle temperature in part (b) but not in part (a). A real gas may give the same result as a perfect gas under special conditions, without obeying the perfect gas law.

2.5. $\qquad \Delta U = \Delta Q + \Delta W; \quad \Delta W = -P(V_2 - V_1)$

$$\Delta Q = C_P(T_2 - T_1) = C_P P(V_2 - V_1)/R.$$

Therefore

$$\Delta U = P(V_2 - V_1)/(\gamma - 1)$$

$$= 7.5 \times 10^5 \, \text{J} \, (C_P - C_V = R)$$

2.6. $\qquad W = -\displaystyle\int_{V_1}^{V_2} P \, dV = -2RT \ln 2 = -6.9 \, \text{kJ}.$

Work done by gas $= 6.9 \, \text{kJ}$. Heat supplied also $= 6.9 \, \text{kJ}$ because $\Delta U = 0$ for a perfect gas at constant temperature.

2.7. $\qquad$ Along ab $W = -RT_1 \ln 2; \quad \Delta U = 0$ Therefore $Q = RT_1 \ln 2$
$\qquad$ Along bc $W = RT_1/2; \qquad Q = -C_P T_1/2$
$\qquad$ Therefore $\Delta U = -C_V T_1/2$

$\qquad$ Along ca $W = 0; \, Q = C_V T_1/2$ Therefore $\Delta U = C_V T_1/2$

Therefore

$$\text{in cycle } W = -RT_1 \ln 2 + RT_1/2 < 0$$

$$Q = RT_1 \ln 2 - RT_1/2 > 0$$

Only U is a function of state since $\Delta U = 0$.

2.8. $W = -P(V_2 - V_1)$ where P is 1 atmosphere, V_2 the volume of 1 mole of vapour at the boiling point and V_1 the volume of 1 mole of liquid. V_1 is negligible relative to V_2.

Therefore $W = -PV_2 = -RT_b = -373R = -3.1 \, \text{kJ}.$

The work done by the vapour is 3.1 kJ. The heat supplied was 41 kJ. Therefore 38 kJ was used to overcome intermolecular forces and only 3 kJ to perform external work.

2.9. $đW_R = \mathscr{F} \, dl$ represents work done on the system. At constant tension $\mathscr{F}_0$

$$W = \mathscr{F}_0(l_2 - l_1) = \mathscr{F}_0 l\alpha(T_2 - T_1)$$

where $\bar{\alpha}$ is the mean coefficient of linear expansion in the temperature range T_1 to T_2.

2.10. If W were a function of state dW would be a perfect differential (Appendix I) and

$$dW = \left(\frac{\partial W}{\partial T}\right)_V dT + \left(\frac{\partial W}{\partial V}\right)_T dV$$

but $\qquad dW = 0\,dT + P\,dV$

and condition for a perfect differential becomes $(\partial P/\partial T)_V = 0$. W cannot therefore in general be a function of state since the pressure and temperature of a system are related by the equation of state.

2.11. $\qquad \delta T = \theta(T_{TP} - T)/(T_{TP} - \theta)$.

2.14. (a) Perfect differential; (b) imperfect differential. Integrating factor $x^m y^{m+1}$.

2.15 $\qquad \Delta Q = A(T_2^4 - T_1^4)/4 + \gamma(T_2^2 - T_1^2)/2$.

2.16. $\qquad c_S = \sqrt{\gamma R T/M}.\,322,\,330\,ms^{-1}$.

3.1.(a) The energy eqn (3.14) leads to the results

(i) $\qquad (\partial U/\partial V)_T = 0$, therefore $U = f(T)$
(ii) $\qquad$ As (i)
(iii) $\qquad (\partial U/\partial V)_T = a/V^2;\ U(2) - U(1) = a\left(\dfrac{1}{V_1} - \dfrac{1}{V_2}\right) + f(T)$
(iv) $\qquad$ As (iii)

In the case of the perfect gas, $U = f(T) = C_V T$ and since the other equations reduce to the perfect gas as V approaches infinity (low density) $f(T)$ has the same form for the other equations.

(b) $\qquad C_P - C_V = T(\partial P/\partial T)_V(\partial V/\partial T)_P$

(using equations 2.34 and 3.14)

The results are therefore

(i) $\qquad R$
(ii) $\qquad R$
(iii) $\qquad R(1 - 2a/RTV)^{-1} \approx R(1 + 2a/RTV)$
(iv) $\qquad R[1 + 2a(V - b)/RTV^2] \approx R(1 + 2a/RTV)$

The repulsive term 'b' has little effect on $C_P - C_V$.
  (c) Use eqn 3.20(d) Put $dS = 0$ in eqn 3.15.
Therefore $C_V\,dT/T = -(\partial P/\partial T)_V\,dV$

Assume C_V is independent of T

(i) $TV^{R/C_V} = \text{constant}$
(ii) $T(V-b)^{R/C_V} = \text{constant}$
(iii) $TV^{R/C_V} = \text{constant}$ $[C_P - C_V \neq R, \text{ see Section (b)}]$
(iv) $T(V-b)^{R/C_V} = \text{constant}$ $[C_P - C_V \neq R, \text{ see Section (b)}]$

3.2. Exercise 2.5: $dS = dQ_R/T = C_P dT/T$ $\Delta S = C_P \ln(T_2/T_1) = (5R/2)\ln 2$
Exercise 2.6: $\Delta U = 0$ $TdS = PdV$ $\Delta S = 2R \ln(V_2/V_1) = 2R \ln 2$
Exercise 2.7: Along ab $\Delta S = R \ln 2$
            Along bc $\Delta S = -C_P \ln 2$
            Along ca $\Delta S = C_V \ln 2$
Total $\Delta S = 0$ since $C_P - C_V = R$ for a perfect gas.
Exercise 2.8: $\Delta S = Q/T_b = 110 \text{ J K}^{-1}$

3.3. See Fig. 3.7
   (a) Heat absorbed along ab = work done = $RT_A \ln(V_b/V_a)$
       Heat rejected along cd = $RT_B \ln(V_c/V_d)$
We now have to relate the volumes using the equation for a reversible adiabatic change

$$T_A V_b^{\gamma-1} = T_B V_c^{\gamma-1}, \; T_B V_d^{\gamma-1} = T_A V_a^{\gamma-1}$$

Therefore

$$V_b/V_a = V_c/V_d$$

Therefore using eqn 3.48; $\eta = (T_A - T_B)/T_A$

   (b) $dQ_R = C_V dT + T(\partial P/\partial T)_V dV = C_V dT + RT(V-b)^{-1} dV$
       along ab $Q_A = RT_A \ln[(V_b - b)/(V_a - b)]$
       along cd $Q_B = RT_B \ln[(V_c - b)/(V_d - b)]$
Now use result of Exercise 3.1(c) part 4. $\eta = (T_A - T_B)/T_A$. The efficiency of a reversible engine is independent of the working substance.

3.4. See Fig. 3.8. The Stirling engine rejects heat reversibly along bc and accepts the same amount of heat along da. The net heat transfer is therefore only along ab and cd

   Along ab: $Q_A = RT_A \ln(V_b/V_a)$
   Along cd: $Q_B = RT_B \ln(V_c/V_d)$
   Also $V_b = V_c; \; V_a = V_d$
   Therefore $\eta = (T_A - T_B)/T_A$

The work done in the Stirling cycle is $R(T_A - T_B)\ln(V_c/V_d)$
The work done in the Carnot cycle is $R(T_A - T_B)\ln(V_c/V_d')$ which is less than that of the Stirling cycle.

3.5. The heat transferred in a reversible process is

$$Q_R = \int T \, dS$$

The area enclosed inside the rectangle (Fig. 3.6) representing the Carnot cycle is therefore equal to the net heat absorbed, just as the area inside a cycle on a P–V diagram is equal to the net work

3.6. Isentropic change $T^3 V = $ constant (see eqn 10.27). Therefore $PV^{4/3}$ = constant along an adiabat and $P = $ constant along an isotherm. Therefore for the isothermal at T_A:

$$Q_A = \int (P+u)\,dV = 4bT_A^4 (V_b - V_a)/3.$$

for the isothermal at T_B:

$$Q_B = 4b\,T_B^4 (V_c - V_d)/3$$

$$V_b T_A^3 = V_c T_B^3; \quad V_d T_B^3 = V_a T_A^3$$

Therefore

$$\eta = (T_A - T_B)/T_A$$

3.7.(a) $\qquad \Delta S_{l(ake)} = Q/T = 500 \ (600 - 300)/300 = 500 \text{ J K}^{-1}$

The entropy change of the block must be calculated by considering a reversible path from 600 K to 300 K. A series of thermal reservoirs at intermediate temperature would form a reversible path and the total change of entropy is

$$\Delta S_{b(lock)} = C \int_{600}^{300} \frac{dT}{T} = -500 \ln 2 \text{ J K}^{-1}$$

The total entropy change is therefore $+155 \text{ J K}^{-1}$.

(b) $\qquad$
$$\begin{aligned}
\Delta S_l &= 500(373 - 300)/300 = 122 \text{ J K}^{-1} \\
\Delta S_{water} &= 500(600 - 373)/373 = 305 \text{ J K}^{-1} \\
\Delta S_b &= -500[\ln(600/373) + \ln(373/300)] \\
&= -346 \text{ J K}^{-1}
\end{aligned}$$

The net entropy change is therefore $+81 \text{ J K}^{-1}$. Cooling in two steps has reduced the net change in entropy. A series of reservoirs would make the change reversible and hence the net entropy change would be zero.

3.12. (i) (a) $n = 0$, (b) $n = \infty$; (c) $n = 1$.

    (ii) $TV^{n-1} = $ constant.

    (vi) $R/2$.

3.14. Otto: $\eta = 0.60$; Carnot: $\eta = 0.80$.

3.15. 1.8.

3.16. 5.0, 0.87.

4.1.
$$P = -(\partial U/\partial V)_S = 2U/3V$$
$$T = (\partial U/\partial S)_V = 2U/3Nk$$

Therefore

$$PV = NkT; \quad U = 3NkT/2; \quad C_V = 3Nk/2; \quad C_P = 5N\overset{\scriptscriptstyle\circ}{k}/2$$
$$F = U - TS = (3NkT/2)\,\{1 - \ln[3T(V/N)^{2/3}/2\alpha]\}$$

Notice that U and F are properly extensive and P and T intensive.

4.2
$$\kappa_T^{-1} = -V(\partial P/\partial V)_T = V(\partial^2 F/\partial V^2)_T$$
$$C_P - C_V = -T(\partial P/\partial T)_V^2/(\partial P/\partial V)_T$$
$$= T(\partial^2 F/\partial T\,\partial V)/(\partial^2 F/\partial V^2)_T$$

4.3
$$C_P = (\mathrm{d}Q_R/\mathrm{d}T)_P = T(\partial S/\partial T)_P = -T(\partial^2 G/\partial T^2)_P$$

4.4. The energy of the system plus the reservoirs must be a minimum. Therefore

$$\mathrm{d}(U + U_R) = \mathrm{d}U - T_R\,\mathrm{d}S - P\,\mathrm{d}V = 0$$

Therefore

$$\mathrm{d}(U - TS + PV) = \mathrm{d}G = 0$$

Since

$$T = T_R, \ P = P_R$$

5.1.
$$A = 3.66 \times 10^{-6}\,\mathrm{J\,mol^{-1}}. \quad C_V = 23.8\,\mathrm{J\,deg^{-1}\,mol^{-1}}$$
$$\text{at } 300\,\mathrm{K}; \ 25.7\,\mathrm{J\,deg^{-1}\,mol^{-1}} \ (800\,\mathrm{K}).$$

Notice that the correction to constant volume at room temperature is only about 3 per cent of the heat capacity.

5.2. Equation 5.13

$$\Delta T = -\frac{a}{C_V}\left(\frac{1}{V_1} - \frac{1}{V_2}\right) = -5.6\,\mathrm{K} \text{ (see Exercise 3.1a)}$$

Only the term representing attractive forces between the molecules occurs in the result for free expansion. Although the cooling appears to be quite large it will in fact be greatly reduced due to the thermal mass of the container.

5.3. We are required to show that

$$(\partial U/\partial P)_T + \partial(PV)/\partial P = V - T(\partial V/\partial T)_P$$

The left-hand side may be written

$$(\partial U/\partial P)_T + P(\partial V/\partial P)_T + V = (\partial U/\partial V)_T(\partial P/\partial V)_T^{-1}$$
$$+ P(\partial V/\partial P)_T + V$$

on substituting for $(\partial U/\partial V)_T$ using the energy equation the identity follows.

5.4.
$$P = \frac{RT}{V-b} - \frac{a}{V^2} \approx \frac{RT}{V} + \frac{RTb - a}{V^2} \quad \text{(using the binomial expansion)}$$

Therefore
$$T(\partial V/\partial T)_P - V \approx (RTb - 2a)/RT$$

Therefore
$$T_i = 2a/Rb$$

This is the maximum inversion temperature (low-pressure limit).

5.5.

$$T\left(\frac{\partial V}{\partial T}\right)_P - V = T\frac{dB_P}{dT} - B_P + \left(T\frac{dC_P}{dT} - C_P\right)P + \left(T\frac{dD_P}{dT} - D_P\right)P^2$$

If the series was terminated with the terms in C_P the inversion pressure would be given by

$$P_i = -(T\,dB_P/dT - B_P)/(T\,dC_P/dT - C_P)$$
$$= -[d(B_P/T)/d(C_P/T)]$$

This equation is adequate at high temperature but does not give a finite maximum value of the pressure on the inversion curve (Fig. 5.2) because C_P/T has a maximum value and at this temperature P_i goes to infinity. Using the van der Waals equation for example

$$C_P = [b^2 - (b - a/RT)^2]/RT \quad \text{(Exercise 2.2)}$$
$$= (2T - T_B)b^2 T_B/RT^3 \quad (T_B = a/Rb = \text{Boyle temperature})$$

Then C_P/T has a maximum value when $T = 2T_B/3$.
(The van der Waals result for C_P is not in good agreement with experiment but the general conclusion is correct. The term D_P becomes important at low temperature.)

5.6. The important parameter in the Joule–Kelvin expansion is $(\partial T/\partial P)_H$ given by eqn 5.19. In a reversible adiabatic expansion $(\partial T/\partial P)_S$ must be considered.

$$(\partial T/\partial P)_S = -(\partial T/\partial S)_P(\partial S/\partial P)_T = T(\partial V/\partial T)_P/C_P$$

Therefore

$$(\partial T/\partial P)_S = \alpha TV/C_P > 0 \text{ (gas cools in isentropic expansion)}$$

$$(\partial T/\partial P)_H - (\partial T/\partial P)_S = -V/C_P < 0$$

The Joule–Kelvin expansion produces a smaller temperature drop than a reversible adiabatic expansion but involves no moving parts, which is an advantage at low temperature.

5.10. T^7.

5.12. 1.37×10^7 Pa (137 at.)

5.13. 3.5×10^4 Pa (0.35 at.)

5.14. -2.1 K

6.1. If the magnitude of v_x is independent of v_y and v_z, the number of molecules with simultaneous components in the range v_x to $v_x + dv_x$, v_y to $v_y + dv_y$, v_z to $v_z + dv_z$ is simply the product function

$$Nf(v_x) f(v_y) f(v_z) \, dv_x \, dv_y \, dv_z$$

This product, by assumption, (a) must be a function of the *magnitude* of the velocity, that is to say of $(v_x^2 + v_y^2 + v_z^2)$

Therefore

$$g(v^2) = g(v_x^2 + v_y^2 + v_z^2) = f(v_x) f(v_y) f(v_z)$$

The solution is of the form

$$f(v_x) = A \exp(Bv_x^2); \; g(v^2) = A^3 \exp(Bv^2)$$

where A and B are constants. A is simply a normalizing factor and B may be found by calculating the pressure exerted by the gas and relating it to the perfect gas law. The weakness of the proof (as Maxwell recognized) is that it cannot be *assumed* that the velocity components are independent. This was first proved by Boltzmann using the H-theorem.

6.2. Substitute eqn 6.1 into 6.3. Then

$$H = \ln\left[\frac{N}{V}\left(\frac{m}{2\pi kT}\right)^{3/2}\right]\int_{-\infty}^{\infty} f(v) \, d^3v - \frac{m}{2kT}\int_{-\infty}^{\infty} v^2 f(v) \, d^3v$$

The first integral is equal to (N/V) and the second to $3NkT/2V$ so

$$S = -kVH = \tfrac{3}{2} Nk \ln T + Nk \ln\left(\frac{V}{N}\right) + \text{constant}$$

in agreement with eqn 3.40 once the identity of C_V and $3Nk/2$ has been established (Exercise 4.1 or Section 7.5).

6.3. From eqn A4.18.

$$\Delta\varepsilon = h^2/2mL^2 = 2.4 \times 10^{-33}\,\text{J} = 1.5 \times 10^{-14}\,\text{eV}$$

(see Appendix IV).
Then

$$T_1 = \Delta\varepsilon/k = 1.7 \times 10^{-10}\,\text{K}$$

The equivalent temperature ($\approx 10^{-14}\,\text{K}$) for a hydrogen molecule is given simply by changing the mass.

6.4. Total energy can only be an integer multiple of ε_0; $E = M\varepsilon_0$ where M is given by the sum of

$$M = n_1 + n_2 + \ldots + n_N = \sum_{r=1}^{N} n_r$$

and $n_r\varepsilon_0$ is the energy of the rth oscillator. The number of distinguishable microstates is the number of ways of bringing about M by changing the n_r. This is equal to the number of ways of distributing M objects into N groups. Now N groups can be formed by using $(N-1)$ partitions between objects. Total number of permutations is now $(M+N-1)!$ but we are not concerned with permutations between objects or partitions so

$$W(E) = (M+N-1)!/M!(N-1)!$$

This type of calculation is discussed further in Chapter 7.
6.5. The probability of a particle being in the state of lowest energy for example is

$$P(0) = \frac{(4 \times 5) + (3 \times 20) + (2 \times 30) + (3 \times 10) + (5 \times 1)}{\text{Total number}}$$

$$= \frac{175}{350} = \frac{1}{2}$$

$$P(0) = 0.50,\ P(\varepsilon) = 0.29,\ P(2\varepsilon) = 0.14,\ P(3\varepsilon) = 0.06,$$

$$P(4\varepsilon) = 0.01$$

The most probable arrangement occurs 30 times and within this arrangement

$$P(0) = 0.40,\ P(\varepsilon) = 0.40,\ P(2\varepsilon) = 0.20,\ P(3\varepsilon) = 0,$$

$$P(4\varepsilon) = 0$$

The agreement between the two distributions is already quite good for five particles and becomes rapidly better as the number of particles increases. An extensive discussion of this type of calculation may be found in J. G. Powles, (1968).

6.6. $\qquad W = (N+n)!/N!n!; \ E = n\varepsilon_0$

Therefore

$$S = k \ln W = k[(N+n)\ln(N+n) - N \ln N - n \ln n]$$

Now *either*

$$(\partial S/\partial E)_V = 1/T = (\partial S/\partial n)_V/\varepsilon_0$$

Then

$$(\partial S/\partial n)_V = \varepsilon_0/T = k[\ln(N+n) - \ln n]$$

$$n = (N+n)\exp(-\beta\varepsilon_0) \approx N \exp(-\beta\varepsilon_0)$$

or $\qquad F = E - TS = n\varepsilon_0 - k[(N+n)\ln(N+n) - N \ln N - n \ln n]$

and, since F is a minimum at thermal equilibrium, $(\partial F/\partial n)_V = 0$. This type of imperfection in a crystal is called a Schottky defect.

6.7. Distinguishable: $g_r^{nr} = 9$ ⠀⠀Bosons: ⠀⠀$4!/2!2! = 6$.
⠀⠀⠀⠀⠀⠀⠀⠀⠀⠀⠀⠀⠀⠀⠀⠀⠀Fermions: ⠀⠀$3!/2!1! = 3$.

6.8. Helium ⠀⠀⠀⠀⠀⠀⠀⠀⠀⠀⠀⠀⠀$e^{-\alpha} \approx 10^{-3} \ll 1$

Sodium one electron per atom. Lattice spacing about 0.3 nm. Hence $e^{-\alpha} \approx 10^6$.

6.10. (a) 0.14, (b) 0.06.

6.15. (d) $h \ll kT/mg$; $h \gg kT/mg$

⠀⠀⠀(e) NkT/Ah; Nmg/A

7.1. ⠀⠀⠀⠀⠀⠀$\Delta/kT = 2\mu_0 H^*/kT = 7 \times 10^{-4}$

for the nucleus. Therefore it is still in the high temperature limit (Exercise 7.2).

⠀⠀⠀⠀⠀⠀⠀$\Delta/kT = 1.4$

for the ion, so the full expression for the thermal capacity is required.

7.2. ⠀⠀⠀⠀⠀$C_V = [Nk\beta^2\varepsilon^2 \exp(\beta\varepsilon)]/[1 + \exp(\beta\varepsilon)]^2$

High-temperature limit:

$$C_V = (N\varepsilon^2/4k)T^{-2} \quad (\beta \to 0)$$

In the helium region (4 K) the thermal capacity may *increase* due to this term while other contributions to the thermal capacity are decreasing (Section 10.3).

7.3. The partition function is given in eqns 7.49 and 7.50. Then

$$F = -kT \ln Z, \quad S = -(\partial F/\partial T)_V = -(\partial F/\partial \beta)_V (d\beta/dT)$$

Therefore

$$S/Nk = \beta\varepsilon_0/[\exp(\beta\varepsilon_0) - 1] - \ln[1 - \exp(-\beta\varepsilon_0)]$$

7.4. Use eqn 7.6 for the partition function and eqn A4.33 for the density of single-particle states. Then

$$Z = \frac{1}{N!}\left[\left(\frac{4\pi V}{h^3 c^3}\right) \int_0^\infty \varepsilon^2 e^{-\beta\varepsilon} \, d\varepsilon\right]^N$$

$$= \frac{1}{N!}\left(\frac{4\pi V}{h^3 c^3}\right)^N [(kT)^3 \Gamma(4)]^N$$

$$F = -kT \ln Z = -NkT [\ln(T^3 V + \text{constant}]$$

$$P = -(\partial F/\partial V)_T = NkT/V, \text{ as for a Newtonian gas but}$$

$$\bar{E} = F - T(\partial F/\partial T)_V = 3NkT = 3PV \text{ (compare with eqn 7.67)}$$

7.5. $\qquad \varepsilon_n = -hR_\infty/n^2$

is the energy of the nth energy level and it is $2n^2$ degenerate. Therefore

$$P_n/P_1 = g_n \exp(-\beta\varepsilon_n)/g_1 \exp(-\beta\varepsilon_1) = n^2 \exp[\beta(\varepsilon_1 - \varepsilon_n)]$$

$$P_4 : P_3 : P_2 : P_1 = 2 \times 10^{-10} : 4 \times 10^{-12} : 2 \times 10^{-10} : 1$$

The upper levels are sparsely populated, even at 5000 K.

7.6. $\qquad P_J/P_0 = (2J + 1)\exp[-J(J+1)h^2/8\pi^2 IkT)$

$$P_4 : P_3 : P_2 : P_1 : P_0 = 3 \times 10^{-2} : 0.2 : 0.9 : 1.6 : 1$$

Notice that more molecules are in the first rotational energy level than in the ground state.

For equal populations

$$7 \exp(-12h^2/8\pi^2 IkT) = 5 \exp(-6h^2/8\pi^2 IkT).$$

Therefore

$$T = 1.5 \times 10^3 \text{ K.}$$

7.7. (a) $H = kT^2(\partial \ln Z/\partial T)_V + kTV(\partial \ln Z/\partial V)_T$
$\qquad G = -kT \ln Z + kTV(\partial \ln Z/\partial V)_T$

(b) $\ln Z = N \ln z$

(c) $\ln Z = \ln(z^N/N!) = N \ln z + N - N \ln N.$

7.9. $dQ_R = -\mathcal{F} dl$

(a) $\alpha_T = \dfrac{1}{L}\left(\dfrac{\partial L}{\partial T}\right)_{\mathcal{F}} = -\dfrac{1\mathcal{F}}{kT^2}\left(\dfrac{L_M^2 - L^2}{L_M L}\right); -\dfrac{1}{T}.$

(b) $C_L = 0$

(c) $C_{\mathcal{F}} = \mathcal{F}\left(\dfrac{\partial L}{\partial T}\right)_{\mathcal{F}} = \mathcal{F} L \alpha_T.$

7.11. (a) $\sqrt{T}$; (b) $\sqrt{1/\tau}$ i.e. signal to noise improves only as the square root of the time taken for a measurement.

8.1. $\qquad\qquad \overline{v^3} = 0$ (eqn 8.11)

$$\overline{v^3} = \int_0^\infty v^5 \exp(-mv^2/2kT)\,dv \Big/ \int_0^\infty v^2 \exp(-mv^2/2kT)\,dv$$

$$= \sqrt{(128 k^3 T^3/\pi m^3)}$$

(Substitute $mv^2/2kT = x$. The integrals then reduce to standard form as in Appendix II.) $\overline{v^3}$ may also be written in terms of $\bar{v}(4kT\bar{v}/m)$.

8.3. The number of molecules striking unit area in unit time is $n\bar{v}/4 = P\bar{v}/4kT$. Number of molecules to cover unit area $\approx 4/\pi d^2$.

$$\text{Time} = 16kT/P\bar{v}\pi d^2 \approx 22 \text{ s}$$

To half-cover the surface at a pressure 10^3 lower, would take $22 \times \frac{1}{2} \times 10^3$ s ≈ 3 h. A pressure of about 10^{-10} torr is therefore necessary if a surface is to remain clean during an experiment.

8.4. The speed distribution function for the beam is of the form of eqn 8.23. Therefore

$$\overline{v^2} = \int_0^\infty v^5 \exp(-mv^2/2kT)\,dv \Big/ \int_0^\infty v^3 \exp(-mv^2/2kT)\,dv$$

$$= 4kT/m$$

$$\bar{\varepsilon} = \overline{mv^2}/2 = 2kT$$

The mean energy of a molecule in the container is $3kT/2$, but fast molecules are more likely to reach the aperture and go into the beam.

8.5. In a time dt the number of molecules which leave the container through a hole of area A is

$$dn_1 = -n\bar{v}_1 A\,dt; \quad n_1 = n_1^0 \exp(-\bar{v}_1 At)$$

Similarly for molecules of type 2:

$$n_2 = n_2^0 \exp(-\bar{v}_2 At)$$

The concentration is

$$n_1/(n_1 + n_2) = n_1^0/[n_1^0 + n_2^0 \exp(\bar{v}_1 - \bar{v}_2) At]$$

complete separation is therefore only possible as t approaches infinity, that is as n_1 approaches zero.

8.6. The number of electrons striking unit area of the surface in unit time with velocity in the x-direction greater than v_0, where $\phi = mv_0^2/2$ is

$$R = n(m/2\pi kT)^{1/2} \int_{v_0}^{\infty} v_x \exp(-mv_x^2/2kT)\,dv_x$$

writing $z = mv_x^2/2kT$ and integrating

$$R = n(kT/2\pi m)^{1/2} \exp(-\phi/kT)$$

This is the first Richardson equation for the number of electrons emitted from a heated filament. It is not quite correct because the electrons must be treated by Fermi–Dirac statistics. The term $T^{1/2}$ then becomes T^2 but in fact the temperature dependence is dominated by the exponential term.

8.7.

	$\sqrt{\overline{v^2}}$	$\bar{v}$	v_{mp}	$\bar{v}$	$\bar{\varepsilon}$	c_s
O_2	483	447	396	0	6.3×10^{-21}	330
H_2	1930	1790	1586	0	6.3×10^{-21}	1,320

(velocity in ms^{-1}; energy in J).

8.8. 1.0×10^4 K

8.9. $\simeq 10^4$ K

8.10. (a) $\sqrt{(3kT/m)}$; (b) $\sqrt{(0.45kT/m)}$; (c) $\sqrt{(1.5k^2T^2)}$ Fluctuations decrease in importance as $\sqrt{N}$.

8.11. $\bar{\varepsilon}/3 = kT/2$. No. $mv_{mp}^2/2 = kT$.

8.12. (a) 2.8×10^{23} cm^{-2} s^{-1}; (b) 2.8×10^{12} cm^{-2} s^{-1}.

8.14. (a) $f(\varepsilon)\,d\varepsilon = $ constant $\varepsilon \exp(-\varepsilon/kT)\,d\varepsilon$; (b) $(1 + \varepsilon_0/kT)\exp(-\varepsilon_0/kT)$.

8.15. (a) 1.4; (b) 1.006

8.16. 1.9×10^{-4} torr (2.5×10^{-2} Pa).

9.1 (a) Using eqn 9.22, with α equal to 2.5 where M is the molecular weight and C_V is equal to $3R/2$ for 1 mole.

$$\kappa = 14.5 \times 10^{-2} \text{ W m}^{-1} \text{ s}^{-1} \text{ deg}^{-1}$$

(b) $\qquad \eta = 0.5 \text{ nm } \bar{v}\lambda = (0.5\lambda PM/RT)(8RT/\pi M)^{1/2}$

Therefore

$$\lambda = 1.8 \times 10^{-7} \text{ m}$$

(c) Using equation 9.6: $d = 2.2 \times 10^{-10}$ m (note that $d \ll \lambda \ll$ dimension of container)

(d) $\lambda \propto T/P$ so using the result of (c)

$$\lambda = 2.2 \times 10^{-10}(760T/273P) \, (P \text{ in torr})$$

Therefore

$$P = 1.5 \times 10^{-2} \text{ torr (300 K)} \text{ and } 1.5 \times 10^{-3} \text{ torr (30 K)}$$

Therefore $\lambda \approx$ dimensions of the system even in a poor vacuum (compare Exercise 8.3.).

9.2. $\qquad \dfrac{\text{Poiseuille flow}}{\text{Free molecular flow}} \, 16\eta RT/a\bar{v}P = 16\eta/a\bar{v}\rho = 8\lambda/a$

The ratio is therefore equal to $8\lambda/a$, which is equal to 16 for a mean free path equal to the diameter. The observed mass flow is therefore an order of magnitude greater than would have been expected on the basis of Poiseuille flow.

9.3. A change of the total mass of molecules of type 1 in the container of volume V leads to a change in pressure since

$$\Delta m = m_1 \Delta N_1 = m_1 \, V\Delta P_1/kT$$

where m_1 is the mass of a molecule of type 1. Then from eqn 9.28

$$\dot{P}_1 = -\alpha P_1/\sqrt{m}$$

where α contains the parameters independent of the nature of the gas. Therefore

$$P_1(t) = P_1(0)\exp(-\alpha t/\sqrt{m}) = P_1(0)\exp(-t/\tau)$$

Therefore

$$\tau = \alpha^{-1}\sqrt{m}$$

The composition of a gas mixture in the system therefore changes with time since τ is proportional to the square root of the mass of a molecule.

9.4. The radiation loss from unit area in unit time is, from Stefan's law

$$R_R = \sigma(T_w^4 - T^4) \approx 4\sigma T^3(T_w - T)$$

using the binomial theorem. The heat loss through the gas is given by eqn 9.31 so the pressure at which $R_R = R_g$ is

$$P_{min} = 8\sigma T^4/\bar{v} = 6.2 \times 10^{-2} \text{ torr}$$

P_{min} is lowered by a factor equal to the emissivity of the wire if it does not behave like a black body.

9.6. 3.6 Å $(3.6 \times 10^{-10} \text{ m})$

9.7. (a) 4×10^{-7} m, (b) 9×10^{-8} m

9.8. 0.01 m

9.9. I $10^3 - 10^2$, II $10^2 - 10^{-3}$, III $10^{-3} - 10^{-5}$, IV $< 10^{-5}$ at.

9.11. 1.2 m^3 s^{-1}; 6.2×10^{-7} torr $(8.3 \times 10^{-5}$ Pa).

10.1. The Joule–Kelvin coefficient for a gas with the equation of state

$$PV = RT + B_P P \text{ is } (\partial T/\partial P)_H = [T(dB_P/dT) - B_P]/C_P$$

Equation 10.8 may be written for the Bose–Einstein gas in the form

$$PV = RT(1 - B_V/V \dots)$$

Hence

$$B_V = -Nh^3/g(2\pi mkT)^{3/2} = B_P \quad \text{(Exercise 2.2)}$$

Therefore

$$(\partial T/\partial P)_H = 5Nh^3/2C_P g(2\pi mkT)^{3/2}$$

The right-hand side is always positive so a Bose–Einstein gas always cools in a Joule–Kelvin expansion.

The expression for the Fermi–Dirac gas is the same apart from a negative sign and therefore always increases in temperature after a Joule–Kelvin expansion. (The values of C_P are different for the two gases since

$$C_P = C_V + T(\partial P/\partial T)_V (\partial V/\partial T)_P \approx C_V + R + 2P \, dB_P/dT$$

and $C_V = 3R/2$.)

The perfect quantum gases therefore behave like classical gases with attractive (Bose–Einstein) or repulsive (Fermi–Dirac) forces between the molecules (compare Exercise 5.4).

10.2.
$$PV = 2\bar{E}/3 = 2N\bar{\varepsilon}/3 \quad \text{(eqn 10.5)}$$

$$\bar{\varepsilon} = 3\varepsilon_F/5 \approx 3 \text{ eV} \quad \text{(eqn 10.15)}$$

Therefore

$$P = \tfrac{2}{5} \varepsilon_F \cdot \frac{N}{V}$$

$$P = (2\varepsilon_F \times \text{volume per electron})/5 = 19 \times 10^4 \text{ atmospheres}$$

10.3. $u = 4\sigma T^4/c$ (eqns 10.24 and 10.25)

$$C_{rad} = 16\sigma T^3 V/c; \; C_{gas} = 3Nk/2 = 3PV/2T \text{ (constant volume)}$$

Therefore

$$C_{rad}/C_{gas} = 32\sigma T^4/3P \approx 6 \times 10^{-10}.$$

10.4. As a function of frequency, using eqn 10.39, ($h\nu_{max} = 2.82kT$). Using $c = \nu\lambda$, $d\nu = -c\,d\lambda/\lambda^2$ and eqn 10.36

$$u(\lambda, T)\,d\lambda = 8\pi hc\,d\lambda/\lambda^5 \left[\exp[\beta hc/\lambda] - 1\right]$$

$(\partial u/\partial\lambda)_T = 0$ for a maximum in the curve.
Therefore

$$hc = 5kT\lambda_{max}[1 - \exp(-hc/kT\lambda)]$$

Therefore

$$hc/kT\lambda_{max} \approx 5 \text{ (actually 4.965)}$$

Therefore

$$\lambda_{max}\nu_{max} \approx 0.6c$$

The difference arises because a system with a constant-frequency bandwidth allows a greater range of wavelengths to pass at low frequency

$$(\Delta\lambda = c\Delta\nu/\nu^2)$$

10.5. The density of states as a function of the magnitude of the wave vector k may be written in the form (eqns A4.11, A4.16, and A4.23)

$$D(k) = \text{constant} \times k^{n-1} \; (n \text{ dimensions})$$

In the low-temperature limit the dispersion relation is simply

$$\nu = c_s/\lambda = c_s k/2\pi$$

where c_s is the velocity of sound.
Therefore

$$D(\nu)\,d\nu = D(k)\,dk = \text{constant} \times \nu^{n-1}\,d\nu$$

Substituting in eqn 10.47

$$\bar{E} = \text{constant} \times \int_0^\infty \nu^n [\exp(\beta h\nu) - 1]^{-1}\,d\nu$$

$$= \text{constant} \times T^{(n+1)} \int_0^\infty x^n [\exp(x) - 1]^{-1}\,dx$$

and the integral is also a constant so

$$\bar{E} \propto T^{(n+1)}; \; C_V \propto T^n \qquad \text{(for } n \text{ dimensions)}$$

10.6. $D(k)\mathrm{d}k = \text{constant} \times k^2 \, \mathrm{d}k$ as before but now

$$D(v) \, \mathrm{d}v = D(k) \, \mathrm{d}k = \text{constant} \times \sqrt{v} \, \mathrm{d}v \qquad \text{(three dimensions)}$$

Using eqn 10.47 and writing $x = \beta h v$

$$\bar{E} = \text{constant} \times T^{5/2} \int_0^{\infty} x^{3/2} [\exp(x) - 1]^{-1} \, \mathrm{d}x$$

Therefore

$$\bar{E} \propto T^{5/2}; \; C_V \propto T^{3/2}$$

This magnetic contribution to the low-temperature specific heat has been observed in ferromagnetic insulators but is obscured by the electronic specific heat (proportional to T) in ferromagnetic metals.

10.7. The calculation is identical with Exercise 10.5 since the form of the dispersion relation is the same. The magnetic specific heat of an antiferromagnet is therefore proportional to T^3 at low temperature.

10.8. Using eqn 10.113

$$S_S = 0.58 N_s k \qquad \text{(when } x = 1\text{)}$$

The lattice entropy

$$S_L = \tfrac{1}{3} 10^{-5} \, NkT^3 \qquad \text{(at temperature } T\text{)}$$

Therefore

$$S_s = S_L \quad \text{(at 14 K)}$$

$$\Delta S_s = -0.2 \, N_s k \Delta x = 0.2 N_s k \Delta T / T_2 \quad \text{(at } x = 1\text{)}$$

where

$$\Delta T = T_f - T_2, \; \Delta S_L = \tfrac{1}{3} 10^{-5} \, Nk(T_f^3 - T_1^3)$$
$$\approx -\tfrac{1}{3} 10^{-5} \, NkT_1^3 \quad \Delta S_s + \Delta S_L = 0$$

Therefore

$$\Delta T / T_2 \approx 10^{-3} T_1^3$$

$$\Delta T = 2T_2 \text{ when } T_1 = 10 \text{ K.} \quad \Delta T = 10^{-3} T_2 \text{ when } T_1 = 1 \text{ K.}$$

10.9. $C_H = T(\partial S/\partial T)_{H^*}$ (by definition)

Therefore

$$C_H = -T(\partial S/\partial H^*)_T (\partial H^*/\partial T)_s$$

$$(\partial H^*/\partial T)_s = (\partial H^*/\partial T)_{\mathcal{M}}$$

(for a simple paramagnet, eqn 10.108)

Therefore
$$C_H = -\mu_0 T (\partial \mathcal{M}/\partial T)_{H^*} (\partial H^*/\partial T)_{\mathcal{M}} \text{ (using eqn 10.106)}$$
Therefore
$$C_H = \mu_0 C H^{*2}/T^2$$

Using eqn 10.110
$$\int \frac{\mathrm{d}T}{T} = -\mu_0 \int \frac{(\partial \mathcal{M}/\partial T)_{H^*}}{C_H} \, \mathrm{d}H^*$$

Therefore
$$\ln(T_2/T_1) = \ln(H_2/H_1)$$
$$H_2/T_2 = H_1/T_1$$

Index